Franz Sales Meyer

A Handbook of Art Smithing

For the Use of Practical Smiths, Designers of Ironwork, Technical and Art Schools,

Architects, etc.

Franz Sales Meyer

A Handbook of Art Smithing
For the Use of Practical Smiths, Designers of Ironwork, Technical and Art Schools, Architects, etc.

ISBN/EAN: 9783744644358

Printed in Europe, USA, Canada, Australia, Japan

Cover: Foto ©Andreas Hilbeck / pixelio.de

More available books at **www.hansebooks.com**

A HANDBOOK
OF
ART SMITHING

FOR THE USE OF

PRACTICAL SMITHS, DESIGNERS OF IRONWORK

TECHNICAL AND ART SCHOOLS, ARCHITECTS, ETC.

BY

FRANZ SALES MEYER

PROFESSOR IN THE SCHOOL OF APPLIED ART AT KARLSRUHE
AUTHOR OF "A HANDBOOK OF ORNAMENT" ETC.

TRANSLATED
FROM THE SECOND AND ENLARGED GERMAN EDITION

WITH AN INTRODUCTION TO THE ENGLISH EDITION

BY

J. STARKIE GARDNER

CONTAINING 214 ILLUSTRATIONS

LONDON
B. T. BATSFORD, 94, HIGH HOLBORN
1896.

CONTENTS.

INTRODUCTION.

I. Concerning the material.

    1. Iron in general 4
    2. Pig-, and cast-iron 6
    3. Steel 8
    4. Wrougt-iron 9
    5. Malleable cast-iron 12
    6. The various kinds of iron used in trade by Artistic Iron-Workers. Bar-iron, Fancy-iron, Sheet-iron, Iron-wire, Iron-tubes 12

II. Tools and workmanship.

    1. Tools and Machinery 19
    2. The Manipulation and Treatment of wrought-iron . . . 36
    3. The ordinary Iron Combinations 40
    4. The Minutiæ and Details which occur most frequently in Artistic Iron-work 44

III. The Historical development of artistic smithing.

    1. The Antique 51
    2. The Mediæval 53
    3. The Renaissance 60
    4. The Baroque 69
    5. The Rococo 73
    6. The Louis XVI and Empire styles 80
    7. The present day 80

IV. The principal spheres of the smith.

    1. Grilles and balustrades 88
    2. Doors and gates 111

CONTENTS.

	page
3. Hinges and Mountings	128
4. Locks and Keys	139
5. Gargoyles, and hanging Signs	146
6. Candelabras, Candlesticks, Chandeliers, Coronas and Lanterns	151
7. Wash stands and Flower stands	167
8. Crosses for Graves and Towers	180
9. Arms and Armour	183
10. All other Objects in iron	194

SUPPLEMENT.

	page
Tables of Weights and Measures	203
a. German sheet-iron scale	203
b. German wire-scale (millimeter-scale)	203
c. Table shewing the dimensions and weights of wrought-iron gas-barrel	204
d. Table of weights of round bar iron	204
e. Table of weights for square bar iron	205
f. Table of weights for flat bar iron per linear meter in kilogrammes	206

INTRODUCTION TO THE ENGLISH EDITION.

Professor Franz Sales Meyer's previous works on the Science of Ornament, of which his "Handbook of Ornament", in the best known in this country, entitle his views and writings to respectful attention, though his knowledge of smith-craft is theoretical rather than practical. This, to one less highly trained, would have proved a serious difficulty, but brought up from his cradle in an atmosphere of technical education, he has made himself thoroughly acquainted with the métier of which he treats. Passing from the Teachers Training College at Meersburg to the Technical Academy at Carlsruhe, he has been successively appointed teacher in the Art Trade School and the Building Trade School, and finally, in 1879 at the age of thirty, Professor at that establishment. He has also produced as joint author, handbooks treating of metal work, cabinet work, carpentering, and painting.

Of these, the present work is, to English craftsmen, undoubtedly the most interesting.

Addressed especially to art workmen and designers, though not confined exclusively to German examples, the work is written from the German standpoint, which differs in many respects from the English. The actual technical operations are of necessity the same, but the tools differ somewhat.* A large part of the book and of the illustrations is devoted to modern German productions and design, and in view of the somewhat severe competition the English smith is experiencing, and must anticipate in the future from his confrères on the Rhine, this section is not without its special interest.

The ironworking arts and crafts have been at all times most earnestly pursued in Germany. The earliest contact of Roman armies with the Teutonic nations found them well equipped with iron for war, and throughout the middle-ages, references to the steel weapons of Cologne, Passau, Innsbruck and other centres of manufacture are scarcely disguised by the quaint spelling. A little later we find

---

* Mr. John J. Holtzapffel has kindly revised this section of the book which must greatly enhance its value to English readers.

great quantities of "Almagne rivetts", bills, &c., entered in the inventories of military stores in this country; and Henry VIII settled German workmen in Southwark and Greenwich, when endeavouring to revive the armourers craft in England. Finally we now know that the most costly suits of armour in the Paris, Madrid, and Vienna collections were produced in Augsburg, Nuremberg and Munich, whose master armourers achieved world-wide celebrity. The most distinguished artists of the day, Wohlgemuth, Holbein, Dürer, Miehich, Schwarz, Hirschvogel, Flötner, Aldegrever, furnished the designs. To the arts of embossing and encrusting armour with precious metals, known to antiquity, these masters added engraving and etching the steel, besides practising those of painting, tinning, and gilding iron known to Theophilus. Probably too, the art of iron casting was re-discovered in Germany, for cannon of the largest calibre were being cast at Erfurt long before the close of the 14th century. The art of drawing wire is also credited to one Rudolf of Nuremberg, who introduced it soon after the year 1300. It is certain that German ironworkers were peculiarly expert and painstaking before, as well as after the Renaissance, and among the marvels they produced, besides the exquisite shields, sword-hilts and pierced horse-muzzles, the beautiful work put into domestic utensils, tools, instruments of torture even, strong boxes, statuettes carved from the solid, and such tours-de-force as the throne presented to Rudolph II by the Augsburgers in 1574, now in this country, are most remarkable.

It is not however only the beauty of the productions themselves that makes German ironwork so peculiarly worthy of study. While in England, France, Spain, Italy, and the Low Countries, the iron industries ebbed and flowed with changing fortunes, so that they were at times in full and active swing, and at other times dormant almost to the verge of extinction: they did not languish in Germany from the 13th century, and enjoyed continued and boundless prosperity without a break, except during the 30 years war, almost until the invasions of the first Napoleon. Nor were the opportunities for development afforded to German iron industries limited to time alone: they had space and the advantages of racial divergencies as well, for it appears that blacksmithing at least was practised ubiquitously in the land from the Rhine to the farthest limits of Prussia and Austria, and from the confines of Denmark to the Italian frontier.

Of the earlier styles of German ironwork we know little. Of the Romanesque doors which have preserved their iron hinges and guards, some resemble in a remarkable manner the rude contemporary work of central France, whilst others imitate the more carefully designed swaged work of Paris. It is only in the 13th century that blacksmithing begins to exhibit any independent characteristics in Germany. At Marburg, Magdeburg, and many other places we meet with rather elegant branching strap-work on 13th century doors, ending in singular little fleur-de-lis and vine leaves still derived, but diverging considerably, from the French. The divergence continued during the next two centuries and resulted in some rich and characteristic foliated ornament, always based on the vine, mingled with fleur-de-lis and tracery forms. After nearly two centuries, and on the eve of the Renaissance, a new style of work appears, at first apparently in Cologne, based on the thistle. The origin of this may safely be assigned to the singular renown achieved by the Matsys family of smiths of Louvain, a specimen of whose work exists in the celebrated Antwerp well-cover. The thistly look of the foliage in this example is well rendered in Fig. 46 and 48, and some of the corresponding German thistle designs face it on page 62, and appear in the lantern, figure 179. Mixed thistle and tracery designs held the field until supplanted by Renaissance ornament.

The Renaissance work is fully illustrated by Prof. Meyer, and it is at this period more especially that Germany presents a perfectly unique field for the study of the ironworkers crafts. It is in the first place to be observed that the development, of blacksmithing at least, was entirely left in the hands of the workmen themselves. Except as designers of armour, artists of note did not meddle with the ironworker, the architects even, giving the smiths a free hand and apparently imposing no conditions as to design. There were no factories, and the nearest approach perhaps to any teaching school was the concert of a master smith with his numerous apprentices and assistants. Of designers of ironwork, as designers there were probably none, the master smith setting the task and directing the work on strictly traditional lines, with such modifications only as the moment suggested. The work may, in most cases, have been produced without drawings, for ironwork designs followed certain definite lines of precedent, which might be modified within limits, but were not departed from. Thus Grilles were often worked from a threadeled centre of more or less complexity, with the loose ends of bars finishing in traditional floriated ornament. Progress was mainly if not wholly confined to increasing the technical difficulties to be overcome by the smith. Not an illustration or drawing of any scrap of blacksmith's work, drawn for its own sake, has come down to us; a fact most remarkable in an age so prolific in studies and designs for the work of the gold and silver smith. Those among us who desire to see this state of things reestablished among the craftsmen of the present day, cannot do better than study attentively the progressive development of German ironworking, from the close of the mediæval period until the style known as Baroque began to change the current of smithing.

The new style came from across the French frontier and spread eventually over a large part, if not the whole of Germany, changing the character of the design and modifying considerably all the traditions of the smith's craft. It was however but a mere wave of fashion compared to the overwhelming change wrought by the Rococo, which followed and swept alway every landmark of the smith. The lilies and passion-flowers, the tricky interlacings, threadles and spirals which had been his peculiar pride, and the round bar itself disappeared at once, only to reappear in our own times. Highly trained professional designers became indispensible, numbers of pattern books were published in imitation of the French, and the smith as creator and designer became extinct. The individual fancy of the workman in Germany could in future only be indulged, if at all, to the most limited extent. The designs were essentially French, but modified in the directions both of extra richness and less restraint. Though the skill and smithcraft in the finer examples is simply superb, the names of the smith's who produced them are never, unless accidentally, recorded. Whether this complete revolution was for good or ill is a debatable question.

29 Albert Embankment
    London, S. E., March, 1896.

J. STARKIE GARDNER.

Introduction.

The universal importance of iron at the present day is beyond all question. Two words "Railway" and "Steam Engine" suffice to prove this. It is impossible to conceive modern life without iron. The plough that tills the land and the weapons that defend it are made of iron. The number of articles formed out of this civilising medium in every imaginable field is incalculable. The armour-plated colossus that cleaves the waves, the Eiffel-tower with which the modern Babylon has surpassed all pre-existing buildings on earth, the death dealing giant-gunnery, impress us on the one hand, while the steel pen, the needle, and the watch-spring, are our indispensable servants on the other.

Iron has been called the proletarian of metals, and this, evidently for the reason that it is to be found everywhere, in worked or unworked state, furthermore, because it is in itself both unimposing and of low value. But on the other hand it has been found that work ennobles this proletarian. It does not, like its more distinguished relative, gold, present itself to the seeker in a pure and perfect state, and only by the employment of a mighty amount of physical and mental power has mankind been able to disassociate it from its normal associates.

Its great powers of resistance its toughness allied to great elasticity and flexibility, and the many-sidedness of its qualities have made it what it now is to us, and have gone so far that its different forms — Steel and Iron — enrich our language with the roots of words.

Iron is in a certain sense a Culture-Gauge. It has been utilised and turned to account during, in round figures, 5000 years; it has risen in importance in proportion to the progress of civilisation, first slowly and then ever quicker and more uninterruptedly, so that the present century has produced far more than all the preceding centuries put together.*)

As figures are accepted as the best evidence, a few may be quoted here, in order to give a general idea of the rôle iron plays in the present age. The production of raw iron for a year in the whole world was estimated (1882) at 21,000,000 Tons, or 420 million Cwts. By refining the raw material we obtain the following increase in value. Whereas a ton of iron in the ore represents about 5 s., a ton of raw iron costs about £ 2.15. — in round figures, while the same weight of unwrought malleable iron already shows a value of about £ 7.10. —**). If the latter is converted into knife blades its value becomes about £ 2000, if into the finest watch-springs its worth will even reach £ 5,000,000, which latter figure represents a million-fold fructification of the raw material.

Iron is of all metals the greatest factor in the composition of our planet. Its ores are found in countless places on the earth's surface. Iron-oxide or combinations of oxygen with iron form a not unimportant element in the rocks forming the earth's crust. The yellow and red colourings in the deposits of loam and clay, chalk and sandstone, are the results of admixtures of iron. When it is considered that the specific weight of the earth's mass is greater than that of the earth's crust and that the increase of heat towards the interior would render, at a limited depth, combinations with oxygen impossible, one is forced to the conclusion that there must be a considerable quantity of iron in an unalloyed condition in the interior of our planet.

That iron also forms a constituent part of other heavenly bodies is proved by the spectral analysis of our sun, and of other central bodies of the universe, such as Sirius and Aldebaran. In confirmation of this conclusion, meteorites fall from time to time out of space on to our planet. These straying visitors, which one may assume

---

*) The increase in the last decade but one may be gathered from the following figures. The out-put of raw-iron in Germany, including Luxemburg, amounted

in 1834	to	110 000 Tons	in 1864	to	905 000 Tons
„ 1844	„	171 000 „	„ 1874	„	1 906 000 „
„ 1854	„	369 000 „	„ 1884	„	3,527 000 „

The production of this metal, therefore, formerly about doubled itself with each decade, while in the course of 50 years it has increased as nearly as possible 32-fold.

**) Germany's production, Luxembourg included, shows the following results:

Year	Iron-ore		Raw-iron		Forged iron	
	Tons	Value	Tons	Value	Tons	Value
1882	8 248 869	£ 1 953 502	3 340 550	£ 9 629 545	1 423 629	£ 10 846 863
1883	8 736 426	£ 1 949 706	3 419 635	£ 9 046 369	1 411 235	£ 10 418 666

to be fragments of fixed stars or of comets, sometimes consist of native iron, and point to homes where the atmosphere must be wanting in oxygen and composed of hydrogen.

If we, after these few remarks about iron in general, turn to the real aim of this manual, the first points to consider are its production and properties, and more particularly those of malleable-iron. These are dealt with in the first section, which is entitled "The technology of the material". The second section deals concisely with the manipulation of malleable iron and the requisite working tools. The third section treats of the historical development of smithing. The fourth, last, and at the same time most comprehensive section, is that relating to the principal applications of art smithing and the chief productions of art smiths arranged under chapters. At the end of the manual are various tables of weights and measures, calculations, &c. Such a supplement hardly seems necessary, nevertheless it will doubtless be welcome to many who find use for a book of this kind. In like manner the list of works relating to iron and artistic smithing may prove welcome to some readers. This catalogue may at the same time be regarded as a concise list of the sources from which this work has been compiled.

SECTION I.

THE TECHNOLOGY OF THE MATERIAL.

1. IRON IN GENERAL.

Iron (German: Eisen, Latin: ferrum) is classed among the so called base metals. Chemically pure iron is an element or base (Fe = 56) which has only a scientific interest. The iron used in arts and manufacture is no more chemically pure than that which is found in nature. The iron ores are, on the average, combinations of iron with oxygen. When this oxygen is expelled by means of heat in coal fires, carbon is absorbed into the metal and technical iron is the result.

The greater or less proportion of carbon is of chief importance in determining the technical quality of iron, whereas other alloys are less useful and in many instances diminish, or even destroy, the utility of the material. Raw, or pig and cast iron contain, speaking generally, the largest proportion of carbon, malleable iron the least, while steel stands in this respect between the two.

The iron-ores to be principally considered are:
1. Magnetic iron-ores (with 72% of iron), of which, the best and most prized is the Swedish iron;
2. Red hematite ores (with 70% of iron). The principal forms of this are iron-glance (Sweden, Lapland and Elba) and red iron ore (Germany, France, England, Spain and Africa);
3. Brown iron-ore (with 50 to 60% of iron). A peculiar form is the so called pea-ore (Luxembourg, Lorraine, Rhineland, Thuringia, Carinthia, Bohemia and Belgium);

4. **Sparry iron-ore** (Carbonate of protoxide of iron). This only yields up to 48% of iron, but is good as a flux (Siegerland, Styria and Thuringia). In kidney and ball-forms it is called **Spherosiderite**. The clay iron-stone and the Blackband are most used in England;
5. **Bog iron-ore.** This is formed by the precipitation of the iron contained in fenny waters. Low lying plains of North Germany, Silesia, Holland, Russia &c.

These ores are widely distributed over the earth's surface. The manufacture of iron is carried on principally by England, North America, Germany, France, Belgium, Austro-Hungary, Russia and Sweden*).

After the iron-ore has been broken up by stamping, crushing, or rolling, sorted and picked by manual labour, and assisted by previous exposure to the weather or by calcination, either in open hearths or in kilns, the mechanical operation of mixing and getting the ores ready for smelting follows. This means the mixing of rich and poor ores together in the right proportions, or in case of need, the admixture of earthy substances (deads) in order to ensure the proper slag which is of the utmost importance in the next operation. Fluor-spar, lime, clay, quartz and marl are the most used fluxes. In addition to the iron ores old iron is used to facilitate the extraction of the raw iron.

The smelting of ores was originally effected in open hearths, or bloomeries by which, however, not raw, but malleable iron or steely malleable iron was produced. Smelting-furnaces came into use about the end of the 15th century, and from these modest beginnings the blast-furnaces now in general use have gradually been developed. The smelting-process in blast-furnaces is effected by introducing a continuous current of hot-blast or heated air (the blast-furnaces remaining in continuous use for from $1^{1}/_{2}$ to 20 years). The ores and combustibles are charged from above through the mouth (also called throat and furnace-top) together with the flux; the smelted iron is run off every 12 to 24 hours from below through the tapping-hole.

In earlier times charcoal alone was used as fuel. When the cheaper coals were introduced, smelting-works were removed from

---

*) The production for the year 1882 gives the following figures:

Great Britain . .	8620000 Tons	Russia . . . . . 463000 Tons
United States . .	4700000 „	Sweden . . . . . 399000 „
Germany	3172000 „	Spain . . . . . 120000 „
France	2033000 „	Italy . . . . . 25000 „
Belgium	717000 „	Other Countries . 102000 „
Austro-Hungary .	530000 „	

the richly wooded districts to the coal basins. It is of the greatest advantage to find deposits of iron-ore and coal together. Coal is converted into coke for use in smelting, in a manner similar to the conversion of wood into charcoal. One Cwt. of raw iron requires in smelting about $1^{3}/_{10}$ Cwt. of coke.

As before mentioned, the greater or less proportion of carbon determines the difference between the three principal sorts of iron. But the limits of variation are not determined by this alone, for other alloys, such as manganese, phosphorus, silica, arsenic and sulphur, greatly affect the quality of iron. Iron may be described as **malleable** when, on being quenched in water, it does not gain materially in hardness and is capable of being **welded**. Malleable and weldable iron becomes **steel** when by tempering it is hardened and strikes sparks from flint. Raw-, pig-, or cast-iron is that which can be neither hammered nor welded.

Until quite recently the various kinds of iron were classed under these three principal heads. However, the progress made in the field of iron production, and the numerous new processes of manufacture have created a number of intermediate and transition forms, so that the old fashioned classifications are no longer pertinent, although they may still remain in use in ordinary parlance. For this reason, before the various kinds of iron are discussed, a table is given on the next page showing the terms now generally adopted and recognised in modern usage.

2. PIG-, AND CAST-IRON.

Pig-, and cast-iron (German: Roh- or Guss-Eisen) bears the first name while in the form of unmanufactured blocks (pigs), and the last when representing manufactured articles. It contains an admixture of carbon of from $2^{3}/_{10}$ to $6\,°/_0$; it melts at from $1050°$ to $1300°$ of Celsius ($= 1858°$ to $2308°$ of Fahrenheit) and is, generally speaking, lighter in weight in proportion to the higher percentage of carbon it contains. Its specific weight is from $6^{7}/_{10}$ to $7^{8}/_{10}$ or an average of $7^{25}/_{100}$. The pressure resisting power is comparatively great, the tension power relatively small. A peculiar property of cast-iron is its **swelling**, which is produced by heating and which remains after cooling.

Cast Iron is classed as hard white or soft grey. Mottled pig-iron is between the two, and is strong or weak mottled according to the preponderance of white- or grey-pig. White-pig is crystalline, brittle and specifically heavy and has a shrinkage of from 2 to $2^{1}/_{2}\,°/_0$. Grey-pig has a granular fracture, is specifically lighter, softer, tougher and better to manipulate than the white-pig; it is more fluid and in consequence fills the mould better in casting.

CLASSIFICATION OF THE VARIOUS KINDS OF IRON.

I. **Non-malleable Iron.** Lighter when smelted; melts suddenly when under heat.		II. **Malleable iron.** Melted with difficulty; softens gradually under heat.	
		A. Steel. Hardens well.	B. Malleable or Wrought Iron. Can scarcely be hardened.
A. White cast iron.	B. Grey cast iron. (Used for castings and called Cast-Iron.)	During preparation is fluid (homogenous). 1. **Cast-Steel** (free from slag). Bessemer Steel Siemens Steel Martin Steel Uchatius Steel Crucible Steel	Fluid under preparation (homogenous). 1. **Cast Iron** (slag-free). Bessemer iron Siemens iron Martin iron Perrnot iron
Carbon chemically associated. (Without graphite.) Spiegeleisen Refined white iron Common white iron	Carbon principally disseminated graphite Silver grey iron Grey Forge	Is doughy during preparation 2. **Weldable Steel** (contains slag). Bloom- or Bessemer Steel Litharge Steel Puddled Steel Cementation Steel Shear Steel	Is dough-like under preparation 2. **Weldable iron** (contains slag). Bloom-iron Litharge iron Puddled iron Welded Packet iron
Mottled pig-iron (in German "Trout-iron", a mixture of white- and grey-pig) Strong mottled (white-pig preponderating) Weak mottled (grey-pig preponderating)			

Its shrinkage averages $1\,^1/_2\,\%$. Grey-pig contains some of its carbon in the form of graphite.

3. STEEL.

Steel (German: Stahl) contains carbon to the extent of from 0.6 to $2.3\,\%$; it melts at from 1300 to 1800° of Celsius (2308 to 3208° Fahr.). Its specific gravity is from 7.4 to 8.0 or an average of 7.7.

Under the original mode of working iron in open hearths or bloomeries and in small furnaces, steel was to a certain extent the result of accident, as the iron was then of a more or less steely character (Bloom-steel). The present methods of producing steel can, in the main, be classified as of three kinds. Firstly steel may be produced by withdrawing a portion of the carbon contained in the melted iron, by means of a blast of air. This is effected either by puddling in hearths or furnaces with a moderate blast under cover of slag (puddled steel), or by an air-blast which is forced through molten iron contained in pear-shaped retorts, this consuming a portion of the carbon and expelling the incombustible impurities and slag (Bessemer steel). In this operation by sampling the slag and also by spectral analysis, the progress of the process is ascertained.

The second kind consists in adding the necessary proportion of carbon to malleable iron (which is notably poorest in carbon) and thus imparting to it the character of steel. For this purpose malleable iron bars are placed in closed boxes filled up with cementing powder (azotic coal; charcoal, horn- and leather-waste, &c.) and kept exposed in roasting ovens at a white heat as long as is necessary to perfect the transformation (cementation steel).

A third method of producing steel is to a certain extent a combination of both of the above processes. Malleable and pig-iron are blended in such manner that steel is the intermediate result.

The circumstance that both the so called shear steel, which is produced from puddled steel by welding, hammering and rolling, and the cementation steel do not give perfect uniformity of texture, has led to the remelting of these kinds of steel into a compact, even, homogenous mass (cast-steel, crucible-steel), the perfecting of which is completed by means of powerful pressure on the still glowing casting by the steam-hammer.

Uchatius-steel is the result of the practical working-out of previous attempts to produce steel from the smelting of iron with iron-oxides. The iron must be granulated for this process.

Martin steel is produced by converting pig-iron, with addition of sparry iron-ore, into malleable iron and melting again with admixture of pig-iron.

The strength or tenacity of steel (absolute, relative and reactive) is great. But the, technically speaking, most important quality of steel consists in its changeability in respect of degrees of hardness, which permits of it being made extremely elastic on the one side, excessively brittle on the other. Red-hot steel that is allowed to cool gradually becomes soft and easily workable, whereas if rapidly cooled it becomes hard, so hard even that it may be powdered. This remarkable material thus admits of being worked, filed, bored, &c. by tools of the same material. Gentle heating (tempering) makes brittle steel elastic or pliant. The degrees of temper are measured by the colour the steel puts on. These grades are as follows:

pale yellow   at 220° C. or 364° F. | uniform purple at 277.6°C. or 466° F.
straw yellow   „ 230 „ „ 372 „ | light blue   „ 288 „ „ 487 „
brown   „ 255 „ „ 427 „ | dark blue   „ 297 „ „ 502.6 „
mottled purple „ 265 „ „ 445 „ | black blue   „ 316 „ „ 536.8 „

If the heat is still more increased this colour scale repeats itself, but less distinctly and in quicker succession. Too frequent heating (over-tempering or burning) makes steel bad and renders it in quality more like malleable iron.

4. MALLEABLE OR SOFT-IRON.

Soft-iron (German: Schmiedeeisen) contains from 0.05 to 0.6 % of carbon. It melts at from 1800 to 2250° Celsius (3208 to 4050° Fahr.). The melting temperatures can only be given approximately, as an exact measurement of such great heat appears to be impossible. For practical purposes malleable iron may be regarded as unmeltable. The specific gravity fluctuates between 7.3 and 8.1, the average being 7.7 to 7.8. Its resisting-power is great in respect of tension (absolute tenacity); that in respect of relative and reactive tenacity is somewhat lower. Compared with steel and cast iron the resisting powers are in kilogrammes (50 kilogr. = 112 lbs English) per square centimetre as follows:

Cast-iron resistance against pressure 7000, against tension 1300.
Steel   „   „   „ 6— to 10000, „   „   6— to 8000.
Malleable-iron   „   „   „   3000,   „   „   4— to 6000.

Malleable-iron is softer than either cast-iron or steel and is most easily wrought; it may be both bent and hammered when cold. It has either a granulated, or fibrous and sinewy texture. Its texture may become changed, and its tenacity reduced, by continuous hammering or concussions. Tempering malleable-iron in cold water renders it only slightly harder. Hammering, drawing and other such-like processes make it harder and more elastic. It becomes soft and weldable again when re-heated and allowed to cool gradually. During heating it passes through various grades as the temperature increases. It becomes first red- and then white-hot, the colour rising from dark red to the most dazzling white. When white-hot malleable-iron becomes so soft that it may be bent, stretched and otherwise worked with the greatest ease; it then becomes weldable, *i. e.* separate pieces of iron may be hammered together into one piece. This weldability is one of the most important qualities for technical purposes. While still glowing, the outer surface of malleable-iron oxidises; iron-scale or hammer-dross forms and falls off; the material loss resulting herefrom is called scale.

As touching the methods of producing malleable-iron, the old open-hearths and bloomeries must first be mentioned. The dough-like lumps of iron produced from the iron ores, in charcoal fires, with the aid of a direct air-blast, were called blooms, and were forged into shape by the hammer. As this process gives but small results and consumes much charcoal, it is very nearly obsolete, and now superseded by the puddling process, *i. e.* converting white or grey pig into malleable iron.

Refining is effected on a refining-hearth. The pig is melted in a charcoal fire aided by a strong blast, through which the drops fall deprived of carbon to coalesce in dough-like lumps below; these are mechanically separated, turned over, &c., and again subjected to the air-blast. The refining process at the same time refines and frees the iron from other impurities. If first class raw material is used, one operation completes the process. With inferior raw material the iron is incompletely refined and must be treated a second time. A third refining may be necessary, the second only making it steel-like, and the third converting it into malleable iron.

The conversion of pig into malleable-iron by puddling is effected in puddling-furnaces by coal-fire; the coals not being allowed to come in contact with the iron in consequence of the sulphur they contain. The raw material is melted in a furnace by a current of intensely heated air and flame, and stirred mechanically through small apertures, or by rotation of the whole vessel, until converted into the dough-like bloom.

The blooms, whether obtained by refining or puddling, are, while intensely heated, squeezed, cut up, welded together, packeted, brought into prismatic form, rolled into bars, &c. and thus purified from slag. The tilt-hammer worked by water power in former times, has of late years been replaced by the far more advantageous steam-hammer. As a rule the refining process gives a purer, closer and tougher iron than is obtained by puddling, while, on the other hand, the puddling operation is cheaper and more resorted to, inasmuch as inferior qualities of raw pig may be treated by it.

Malleable-iron produced by puddling has of late met with severe competition from iron obtained by the Bessemer process, that is by forcing air through molten cast-iron to burn out the carbon, as in the production of steel.

Through refining, or repeated re-heatings, hammerings and rollings the material becomes tough and flexible and the granulated texture changes into one that is fibrous and interwoven.

By running the iron thus obtained through different rolling-mills it becomes closer grained and takes the forms required in commerce, such as, bar-iron, sheet-iron, iron-wire or tubings. As in rolling the widely set rollers are gradually followed by others running closer and closer together, so also in wire-drawing the material is drawn through conical holes in steel-plates, such holes gradually diminishing in size. Oft-repeated drawings necessitate pauses in order to anneal and scour the metal. As malleable-iron takes so many forms, according to the requirements of trade and of workshops, it is discussed more specifically in a later part of this work, so that the general outline here given may suffice for the present.

Malleable-iron as used in trade may, as will have been gathered from the foregoing, be of very different qualities. This depends on the nature of the raw material, the method of preparation and the proportion of impurities. Iron is classified as, **soft and fibrous, soft and brittle** on the one side, and as **hard and tough, hard and inflexible** and **hard and brittle** on the other.

Good malleable-iron must possess the following characteristics: when broken it must show a **dull gloss if of light colour,** and **a bright gloss if of a dark colour.** If the break is white and shining, or grey and dull, it is indicative of a low quality. It must not have been overheated and burnt, must have an equal structure and be free from slag, flaws, cavities, and other imperfections. Forged iron shows, under otherwise equal circumstances, a more granulated break, rolled-iron a more fibrous one. Rolled-iron should show externally a blue-grey to black-grey tint, as a red tinge points to cold rolling and a low degree of tenacity. Forged iron, on the other hand, is nearly always reddish, because it is treated at lower temperatures.

The most salient faults in malleable-iron are flaws, cinderholes, scabs and scales (produced in rolling), longitudinal cracks (from imperfect welding), defective edges (produced in rolling), veins (spots of unequal hardness), cold-short-iron and red-sear (due to too much silicon, phosphorus and sulphur).

Besides the inspection of the broken parts the following tests are also used: 1. dropping the iron-bar which is to be tried from a given height on to an edged block or anvil, when no break must occur; 2. dropping a weight on to the suspended bar; 3. bending a firmly fixed bar backwards and forwards until it breaks; the number of bendings required to break the bar determining the quality, hard iron crackles in breaking, soft iron does not; 4. hammering the iron out while red hot, which must, if of good quality, give a knife-like edge; 5. filing the iron bright and biting it with diluted acid whereupon the veins and cracks become plainly visible. Furthermore, the degree of tenacity may also be ascertained by weighting, and also by an examination of the uncut ends of the bar. Practical smiths rely mostly on the "right feel", which is sometimes satisfactory, but as often the reverse. The sizes and correctness of the desired sections are determined with the caliper or the standard gauge, &c.

5. MALLEABLE CAST-IRON.

Whereas by the refining process, and by puddling, pig iron is decarbonised while in a molten condition, a similar result with castings can be obtained by means of the malleable cast-iron process, which is as follows: small pieces of casting composed of mottled cast-iron with the addition of malleable-iron, are slowly heated to redness and cooled in cast or wrought-iron chests filled with iron oxide (pure red hematite powder, rust, or iron-scale is generally used), whereby a portion of the carbon is absorbed by the oxygen. A physical change seems to go hand in hand with the chemical one, in a manner similar to that which in the tempering of steel reduces its brittleness. Decarbonised castings, finished articles, balusters, mountings, &c. become, as it were, intermediate between cast and malleable-iron and may be treated accordingly. The process is not new, although it is only of late years that it has come generally into use. The change in character of the metal is greatest at the surface and does not penetrate far, so that only objects of not too great thickness can be usefully subjected to it.

As locksmith's work and mountings (iron door furniture), fancy spikes and such-like embellishments for railings, balustrades, &c. are principally subjected to the above process, and such-like malleable-iron castings have long been known to the smith, they must find mention here.

6. MERCHANTABLE IRON USED BY THE SMITH.

Malleable-iron comes to market in the shape of bars of round, rectangular or fancy section, sheets, wire and tubings. The kinds most used in smithing are bar- and sectioned-iron. These may be in charcoal-iron or ordinary iron, forged or rolled. These are again classified according to their subsequent uses into bars for rivets, for railings, wheel-tire-iron, &c.; or, according to measure and weight, into Fine and Coarse iron, or, according to quality, into Extra-quality, Merchantable iron, &c. The particular names used are however chiefly determined by the cross-sections.

A word with reference to the production of these kinds of iron is however in place here. Whereas Square- and Flat-iron (bars and sheets) are not only produced by rolling but also by forging (Bar-, Slit- and Nail-rod iron) or through the slitting of plates (Edge-tool iron), fancy sections are produced exclusively in rolling-mills. These mills, speaking generally, consist of cast-iron or steel rollers which are set in contrary revolution by means of powerful machinery. The rollers lie with their axes parallel in iron frames or beds so that a space exists between them proportioned to the "section" to be produced. Ordinary plain cylinders are used for plate production, whereas for section work the upper surfaces of these rollers are grooved to the necessary forms. When the glowing mass of metal is introduced into the opening it is seized by the rollers, pushed through and impressed to the section. But, as one passage through the mill is generally not enough, the bar has to pass again as often as necessary and be reheated if it has cooled. As the change of form from the plain bar to the perfect section is a gradual process, there are a number of rollers, each with a more developed section than the other, through which the iron passes in turn. Where one pair of rolls will not suffice, two or more frames are united to form a "train". Reversing rolling-mills are such as run both backwards and forwards so that the work is carried on alternately from either end. In ordinary rolling-mills the work must be brought back again to the only available side. As the rollers in the reversible mills run first in one direction and then in the other, the reversing gear is in regular use. The most complete are the three high trains, in which three rollers are placed vertically above each other in the same housings. The lower and middle rollers draw the iron in one direction, while the middle and upper-ones work it back again; this saves time, besides having other advantages.

Round bars (G.: Rundeisen) are circular in section. They are made in thicknesses of from 5 millimetres or about $^3/_{16}$ of an inch upwards. The diameters increase by millimetres up to 30 mm, and by 2 millimetres between 30 and 80 mm, and beyond 80 mm by 5 mm gradation. The principal defects are imperfect roundness and striæ on the surface (mostly on spots opposite to each other as the consequences of flaws in the rollers).

Square bars (G.: Quadrateisen) are square in section. The sizes are similar to those of round iron. The defects are: faulty shape, warped or twisted lengths, sunken sides, blunt angles, striæ and projections milled in on the surfaces and especially on the edges.

Flat-iron (G.: Flacheisen) comprises in its broad sense every right angled section; in its narrow sense it only applies to thicknesses of not less than 3 mm, an eighth of an inch, and not more than

150 mm, 6 inches, broad. When less than 3 mm in thickness it is called

Hoop-iron (G.: Bandeisen). The faults of flat- and hoop-iron are similar to those of round and square iron, but less frequent. The thicknesses of hoop-iron increase from $1/4$ by $1/4$ mm; 00.1 of an inch; those of flat-iron increase by millimetres. The widths increase at first by one millimetre, then by 2, and lastly by 5 mm. Up to the present time there is no universally adopted scale of sizes. Hoop-iron is often described as $1\,1/4$, $1\,1/2$, $1\,3/4$, 2-fold, meaning that the width is 10, $12\,1/2$, 15, $17\,1/2$ or 20 times its thickness.

When the breadth exceeds six inches this iron is classed as sheet iron.

Under **Fancy-iron** is classed all other bar-iron which shows a definite profile in section, and serves for specific trade purposes. There are many forms of it, only a few of which are mentioned here, as they are used exceptionally in smithing, namely: hexagonal and octagonal (a, b, c see Fig. 1), the quarter, half and three-quarter round (d, e, f), hollow half-round, oval- and half-oval iron (g, h, i), casement (k), cover-joint iron (l, m), channel iron (n), equal and unequal angle-iron (o, p), short and long T-iron (q, r), double-T or H-iron (s), U-iron (t), +-iron (u) and quadrant-iron (v), which last-named is much used in America.

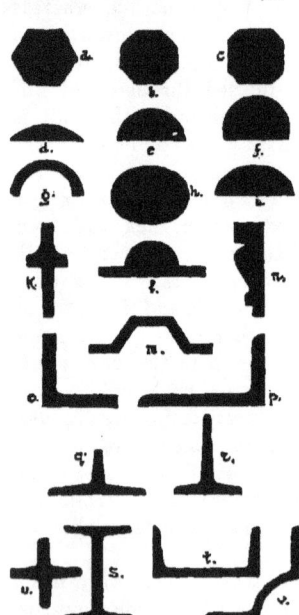

Fig. 1.
Cross Sections of Fancy-iron.

Fancy-iron is classed as ornamental and constructive (builders' iron) according to its kind and use. As touching the latter, so called standard sections have recently been adopted.

The lighter kinds of iron in common use, such as rod- and hoop-iron, are put up into bundles in definite lengths, while heavier sorts and Fancy-iron are sold per bar and by weight.

The prices for the various kinds of iron mentioned, are so arranged that to the fluctuating basis or minimum price are added percentages in permanent ratio for smaller sections, pattern, better quality &c. These extra prices vary according to the manufacturing districts and individual iron-works; it would lead too far to discuss this subject in detail.

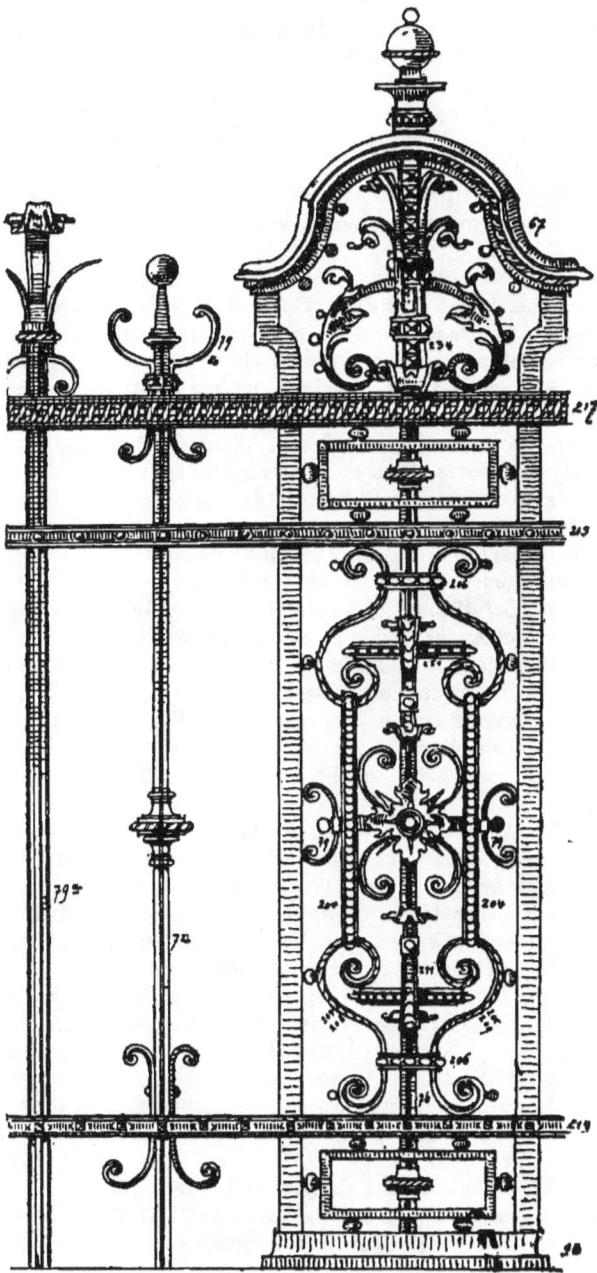

Fig. 2. Railing made of Mannstaedt-Iron.

Forged iron is only preferred to rolled iron at present for certain special purposes. Only Austro-Hungary now supplies forged iron in any considerable quantity,

The **sources of supply** are distinguished according to kind, object and district. It is usual in this respect to designate iron according to the district where it is produced, for instance, Lorraine, Westphalian, Styrian, Upper-Silesian Iron, or else specifically after the name of the works or their owner, such as Burbach Iron, Stumm's Iron.

Of late years the Rolling-mills of L. Mannstaedt & Co. of Kalk, near Cologne, produce various kinds of ornamental sections that are of no small importance to the smith. The handsomely profiled bars whether smooth, or decorated with the Vitruvian scroll twists, foliations, &c. are rolled while glowing, and constitute Fancy-iron of the higher sort. The manufacture does not, of course, permit of undercutting in patterns, but it gives a sharper relief than casting. These bars may, moreover, be bent, twisted and split up as required, so that Mannstaedt's Ornamental Iron is in every respect preferable to the cast-iron strips formerly in use, the more so as the cost is moderate. Fig. 2 shows part of a railing made with the enriched bars referred to, from a design by H. Seeling.

Sheet-iron (G.: Bleche) is either forged or rolled. The forging or rolling from blooms or slabs is either performed singly or else several layers, separated by a surfacing of loam, are worked simultaneously. After the sheets are cut, annealed, cleaned and smoothed, they come to market for sale singly or in bundles. For ordinary dimensions and qualities standard prices are charged; for sheets of special dimensions, larger sizes, better qualities or given fashion, extra-prices are added.

Sheets are classed according to their thickness as heavy, medium or fine. Under the first are numbered armour plating, ship sheathings, boiler plates, tank plates; the medium and fine sheets are either black, or white when coated with tin, zinc or lead. They are often known, not by their thicknesses in mm, but by marks, or the numbers of a particular centre of production. The German Standard Gauge has 26 numbers. No. 1 is 5.5 mm thick, while No. 26 is only 0.375 mm thick. The intermediate thicknesses do not increase in regular gradation, smaller additions being made to the thinner than to the thicker kinds. (See Table in Supplement.)

Besides the ordinary sheets (lock-plates, roofing-sheets, sheets for pipes, &c.) probably only the perforated sheets are used in art-smithing. These are used partly for technical ends and partly for decorative purposes, some of the designs being very pleasing.

The defects found in sheet-iron are scales, bubbles, cinders, cracks and double-plates (the last in consequence of incomplete weldings). Imperfections are detected by tapping with a hammer, when a dull sound is emitted. The quality may also be tested by bending, on which point it must be mentioned that the tenacity is greater in the direction of the rolling than transversely.

Iron-wire (G.: Eisendraht) and steel-wire (G.: Stahldraht) is produced, like sheets, either from malleable-iron or steel. It is first rolled and then drawn through holes bored in steel-plates; the holes being conical and diminishing gradually in diameter; and then reeled, like cables, into coils. The wire is bright, if not annealed after the last drawing, and is more elastic than the dull or black kind. Wire is also often tinned, zinced, coppered, or nickeled. Faultless wire must show a regular cross section; it must have no scale on the surface and no reft inside. The ordinary profile of wire is round. Other sections are made for special purposes. It may here be mentioned that the Chinese and Japanese have a preference for rectangular sections.

The various sorts of wire are often named after the purposes for which they are used, thus: nail wire, field wire, pianoforte wire, shot wire, horticultural wire; or else according to familiar or technical terms, such as: chain rope and hoop wire, or Malgen & Memel wire, or 1st binding wire, 2nd binding wire, &c.; or according to the numbers of one of the centres of manufacture. The German Gauge Scale has 100 Nos besides some further intermediate Nos. Dividing the Wire No. by 10 gives the thickness of the wire in mm. No. 100 has consequently a gauge of 10 mm, while No. 24 is only 0.24 mm thick. Only the thicker kinds of wire are used in art smithing.

Tubings are either cast, which is the case principally with those of considerable dimensions, or rolled out of malleable iron or steel. Under the latter operation the smaller diameters are welded together with butted edges, whereas in the larger sizes the edges are overlapped and welded. This manufacture is principally carried on in England and Germany. The price list is generally a fixed one; the market fluctuations are shown by periodical discount lists. The various articles used in joining tubings such as sockets, socket-ends, L and T joints and bends are supplied with the tubings. The gradations in size increase by $1/8$th of an inch at a time (English), the measurement being taken inside. The comparison between the English scale and millimetres is shown in a special table in the Supplement.

The tubings are made exclusively for technical purposes, such as gas and water-works and fittings, boiler tubes, &c., but artistic smithing often finds use for them, as with chandeliers, brackets, railings, &c. The manufacture of tubings is on the eve of an important revolution. The Brothers Mannesmann have succeeded through the invention of conic rolling-machinery in converting bar iron direct into tubing. As only the best material can be used in this operation, and as the welding is done away with, the Mannesmann tubing will certainly drive the kinds of tubing hitherto in use out of the market. This special tubing is also of importance to smithing inasmuch as rosettes and tendrils can be produced by splitting, bending, &c. Experiments have been made in Munich in this direction with surprisingly favourable results. (See the Journal of the Bavarian Art Trades Association, Year 1892, p. 13.)

The remarks contained in this present section, if taken in connexion with the Tables given in the Supplement, may perhaps suffice to enable the builder's smith, as also the designer, and others to obtain an accurate knowledge of these materials, and also serve to furnish the layman with the information he seeks. We may therefore pass to the second section, wherein the tools used and the manipulation of the materials come under discussion.

SECTION II.

WORKING TOOLS AND MANIPULATION.

1. WORKING TOOLS AND MACHINERY.

Before proceeding to consider the working tools and machinery it is desirable to make a few remarks about the most important appliances necessary to the carrying out of the art smith's craft.

These are the **Hearth** or **Forge**, together with the requisite tools. The forge is built of bricks or made of iron. It is an open hearth with a fire-pit in which the fuel (charcoal, coal or coke) is put. Above the hearth there is usually a projecting sheet iron hood leading to a flue to receive and carry off the smoke and gases evolved. In order to produce and maintain a fierce fire, bellows are fixed so as to introduce the blast either from the side or from below. The bellows were formerly similar to those in domestic use, but they have been of late years greatly superseded by air blasts worked from an engine. The bellows, made of wood and leather, are called pointed, parallel or cylindrical according to their form. They generally consist of two parts, namely the suction-bellows and the regulator, so that they are, in fact, double or compensation bellows capable of giving an uninterrupted blast. This latter is however obtained better by a less cumbersome fan-wheel or some similar arrangement which, like the bellows, can be set in motion by the foot, hand or machinery. In the front part of the forge are found, as a rule, a quenching trough, hollows and receivers for fuel and slack. As belonging direct to the forge itself must be further mentioned the sprinkler or brush, poker, the shovel, and the fire-hook which are respectively used to damp, feed and rake the fire.

SECTION II.

The tools and appliances to be discussed vary greatly according to the work to be executed. First come those used in measuring, and setting out work. Next the supports and means of holding the objects securely; then the various kinds of hammers used in forging, welding, &c.; the tools for cutting and dividing; the drills, borers, punches, the screwing-tools, and lastly all that appertains to

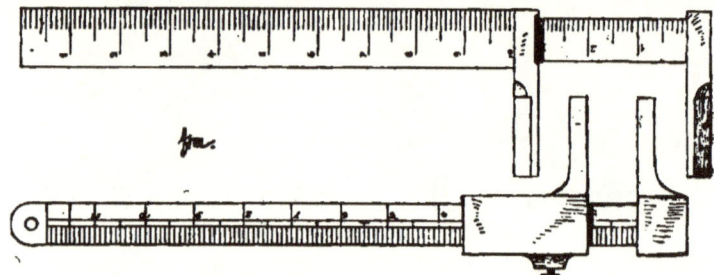

Fig. 3. Examples of Slide-gauges.

cold-working and surface finishing. The most important of these tools will be discussed shortly in the order above mentioned.

a. Apparatus for measuring, setting out, &c.

The **Measures for length** are similar to those in general use, namely: wooden or metal rods or rules, folding, or tape measures.

The **Slide-gauge** used in measuring thicknesses and to deter-

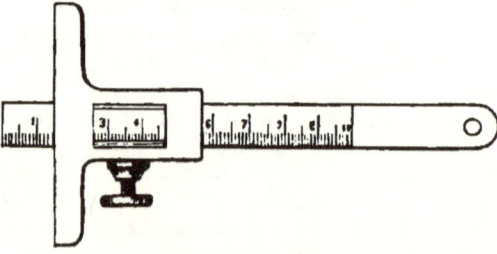

Fig. 4. Set gauge.

mine bulk and capacity is shown in Fig. 3. The socket is generally of brass, the slide and cheeks being of steel. If the socket is scaled as well as the slide the tool can also be used for minute measurements. For more exact measurements the gauge is made with a vernier.

WORKING TOOLS AND MANIPULATION.

The **Set gauge** serves to ascertain the depth of drillings, &c. A plain one with regulating screw is shown in Fig. 4.

Wire- and Sheet-gauges are oblong plates of steel with square or rounded corners having notches along their edges, which are made according to a given scale of dimensions, increasing gradually in size. By inserting the wire or sheet into such gauge the thickness of the article is at least approximately ascertained (see Fig. 5).

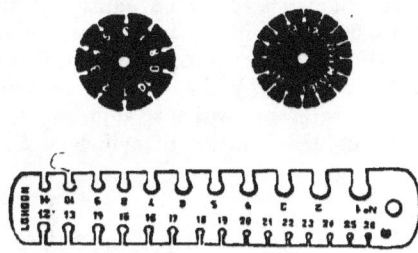

Fig. 5. Wire, Sheet, or Hoop iron gauges.

All plates, &c. with notches along their edges, or elsewhere, which serve to measure standard dimensions and forms are classed as gauges. To these belong therefore the locksmith's gauges which serve to determine the thickness and sections of the wards, size of the shaft or barrel of a key, &c. As a substitute for such gauges lead or wax is often used.

Gauge-pins and -rings (cylindrical rods and hollow cylinders of steel) serve to fix or define the diameters of holes, sizes of cylinders, &c. These are either specially named according to their uses, or generalised as calibre gauges.

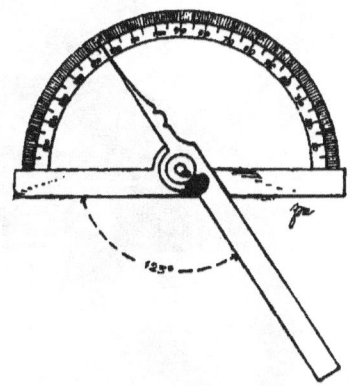

Fig. 6. Goniometer or angle measurer.

Various instruments serve to measure angles. The most commonly recurring right-angle is gauged by a simple iron or steel angle made out of one piece of metal, one of its branches generally being longer than the other. For the reduction and measurement of angles of 30°, 45° and 60° a right angled-triangle of which the angles of the hypothenuse re-

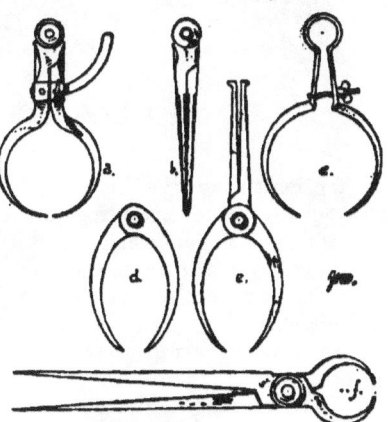

Fig. 7. Various compasses or calipers.

SECTION II.

present 30° and 60°, or 45° and 45°. For measuring other angles the instrument shown in Fig. 6 is useful. The indicator works on a pin and is fixed by an adjusting screw. The number of degrees in the desired angle is read from the scale.

Of the many **Compasses** and **Calipers** in use the most common are: (see Fig. 7) the ordinary calipers and compasses (*a*, *b* and *c*) with or without adjusting arrangement; the calipers for measuring the diameter of cylinders, &c. (*d*); the same combined with

Fig. 8. Plane table scribing block.

legs for measuring the insides (*e*) (by keeping the points at both ends equally wide apart measurements can be obtained in positions from which it is impossible to withdraw calipers while open); and the compasses for determining round and cylindrical dimensions (*f*), one end of which fixes the diameter and the other the circumference.

In transferring, pricking, or drawing the following tools are used:

The **Drawing-table,** a thick right-angled iron plate which must be absolutely smooth and level because it serves as a base (see Fig. 8)

WORKING TOOLS AND MANIPULATION. 23

The **Centre-punches**. These are small punches with conical points which serve to mark the lines to be followed with dots (see Fig. 9).

The **Drawing-needle**, a slim steel pencil for dotting and tracing lines. Brass points are sometimes used.

The **T-Squares, Set-squares**, &c., made of iron or steel.

The **Parallel-rule**, an instrument made in various forms to facilitate drawing lines parallel with the drawing table.

To determine the centre of a circle (which is essential with work that has to be turned on a lathe) the following are used, namely:

the **Centre punch**, an ordinary centre punch moving in a cylindrical box with the end conically widened (see Fig. 10, *a* and *b*), and

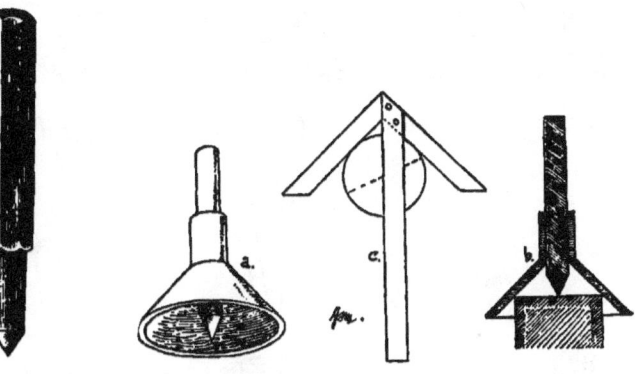

Fig. 9.   Fig. 10.
Centre punch.   Instruments for centering cylindrical objects.

the **Set angle** by which 2 straight lines may be ruled in any direction across the circle, the intersecting point of which will be its centre.

In the fitting-up of machinery, &c. the perpendicular and horizontal plane are abtained by the use of:

the **Plummet**, an elongated metal knob finishing in a point and attached to a string;

the **Plumb-line**, a familiar tool, consisting of an equi-lateral triangle with plummet; and

the **Spirit-level** known to us in its square and tubular forms.

b. Blocks, anvils and gripping tools.

The **Anvil**, which is made of wrought-iron, serves for bearing or supporting the material to be wrought. Its upper face consists of a welded steel-plate which is smooth and slightly domed. The sizes vary. The anvil is fixed either to a wooden-block or else set in a cask filled with pressed sand. Smaller ones are fastened to the bench or remain unfixed. There are beakless, one-beaked and two-beaked anvils (Fig. 11, *a, b, c*). The beaks are conical continuations of the face or hammering surface and assist in forging rings, curves, &c. Anvils of smaller and mostly square faces with 2 long beaks are sometimes called beak-irons or bickerns (Fig. 11, *d* and *e*). Those of cubic form are called stakes (Fig. 11, *f*). Anvils often have holes in the face to receive tools or swages for moulding iron in relief.

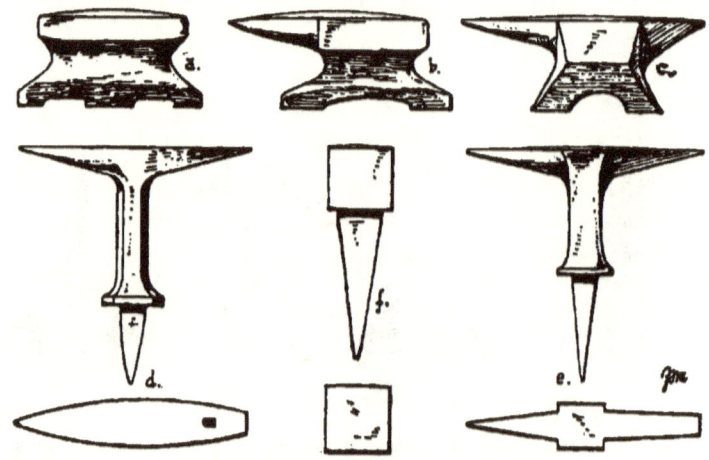

Fig. 11. Various Anvils.

Swages are used when the object to be produced is too difficult or complicated for ordinary forging on the anvil, as in making three-sided and half-round rods, rounded bodies, grooved forms, and ornaments in relief of various kinds. Half-round and 3-sided rods, rods and plates ornamented on one side only, half-round bodies, and such like, only require one swage, which being made of wrought iron, is steeled on the face and set in the anvil. The iron, which has been roughly worked into an approximate form beforehand, is hammered into this swage while red-hot. Whole round forms, bosses, six- and eight-sided rods, &c., necessitate a pair of swages or top and bottom tools, which together may form a closed box or tubular shape. The upper swage is generally shaped like a set-hammer. The articles being forged are either kept in one position, or turned about in the swage, pushed lengthwise forward, &c., according to their character. The swages are made by filing and turning, or by introducing a steel core between the red-hot upper and lower swages

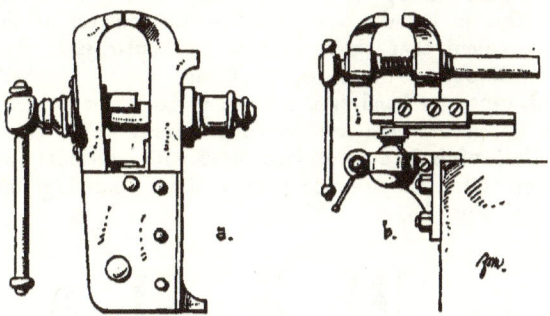

Fig. 12. Vices.

which take its form under the hammer. In order that top and bottom tools may fit properly and not be displaced during the work they are either secured by grooves or held by a spring.

In order to hold the metal fast the **Vice** is required, made in various forms and sizes. It has two cheeks or jaws, one is fixed to the bench or to a special support; the other is connected with it and adjustable.

The jaws are opened and shut by means of a horizontal screw which is set in motion by the iron pin. A spring keeps the jaws open. In the smith's vice (Fig. 12, *a*) the moveable jaw describes a curve, consequently the inner jaw-surfaces are only parallel at one particular width. This disadvantage

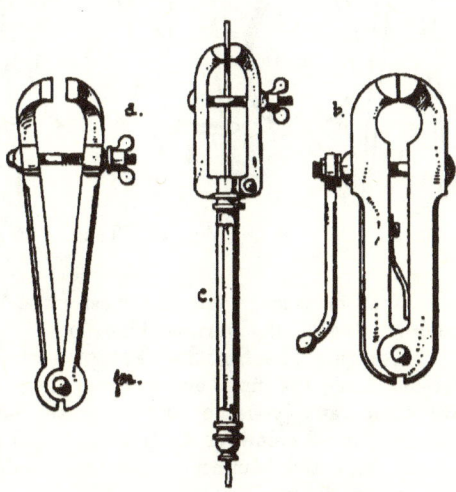

Fig. 13. Hand- and tail-vices.

has brought about the parallel vice (see Fig. 12, *b*) which is made according to various systems. Further kinds of vices are:

False vice jaws or **clamps**. These are made of iron, lead, wood, &c. and are sometimes held together by a spring. They fit between the jaws of the vice, their use being to protect the object held from indentations, &c.;

a **Chamfer-clamp** is a hand-vice, the jaws of which rise obliquely. This in certain work, such as the removal of sharp edges, is more convenient than those with perpendicular jaws.

For small objects use is made of the:

Hand-vice, for hand use, a small unfixed vice, which is opened and closed by means of a key or flanged screw (Fig. 13, *a*, *b*).

The **Tail-vice** is a hand-vice fitted with a handle, which latter is sometimes hollow in order to permit of manipulating long articles, wire, &c. (see Fig. 13, *c*).

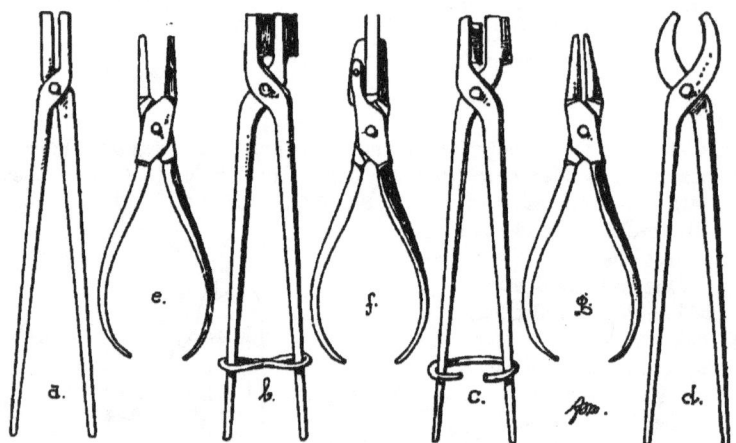

Fig. 14. Tongs and pliers.

Next to vices, come the most important group of shop tools for holding objects, the various kinds of:

Tongs. The **Smiths' Tongs** serve to introduce and withdraw articles from the fire, and to hold them while being forged. They are comparatively large and have the ordinary form, or the jaws may be curved sideways &c. (Fig. 14, *a*, *b*, *c*, *d*). Closed or open rings, driven with the hammer on to the shanks of the tongs, lighten the task of continuous gripping.

The smaller **Flat-nosed pliers** used especially in the manipulation of cold metal, have straight roughened jaws and bent shanks (Fig. 14, *e*).

The jaws of **Parallel pliers** remain parallel to each other whether open or closed (Fig. 14, *f*).

WORKING TOOLS AND MANIPULATION. 27

Wire- or **Round-nosed pliers** have conical jaws and serve, inter alia, to bend or twist wire (Fig. 14, *g*).

c. *Various kinds of hammers.*

Leaving aside the crank-hammer, which is worked with the foot, and the stamp- and tilt-hammer which are worked by water- or steam-power and which (in spite of their advantages) have not come into general use, we still find an exceedingly large variety in the form of

Hand-hammers. These are made of wrought-iron and generally have two hardened steel surfaces, a hole at their centre of gravity

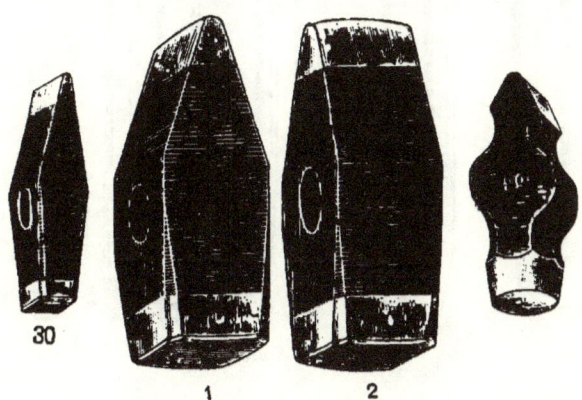

Fig. 15. Smith's or hand-hammers.

and a handle of white thorn or some other tough wood of a round or elliptical section. If the steel surface of the hammer is broad, either circular, or quadrate at both ends it is called a face; if it is narrow, angular or oval at one end it is a "pane". Hammers are also distinguished according to sizes, thus:

the **Sledge-hammer** is from 6 to 20 lbs. in weight, has a handle from 30 to 40 inches long and is used or swung with both hands;

the **Smith's hand-hammer** is from 2 to 5 lbs. in weight and has a handle 12 to 16 inches long;

the **Engineer's hammer** weighs up to 1 lb. and has a proportionate handle.

The **cross pane ordinary Hammers** have a face and a pane at right-angles to the handle (Fig. 15, 1; 30 and 99).

In **Straight panes** the pane runs axe-ways parallel with the direction of the handle (Fig. 15, 2).

The **Block hammer** has two slightly domed faces.

The **Flat hammer** has two flat faces.

The **Hollowing hammer** has two rounded convex faces, &c.

Set-hammers are not actual hammers as they are not used for direct striking. Their forms are those of hammers. They are held in loose handles and serve, like chisels and punches, to cut, round off, make holes, &c. Set-hammers are distinguished as straight, oblique, round, &c.; and as cold-chisels with handles, handled chisels, hollow chisels, chamfering hammers, and such like.

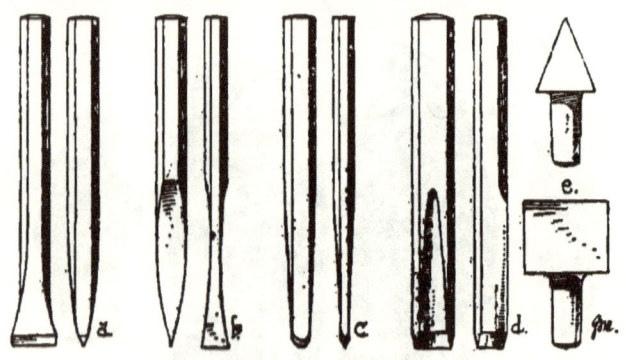

Fig. 16. Chisels.

d. Cutting tools.

Besides the Set-hammers other tools are used for cutting and chiselling. These are:

Cold- or **cutting-chisels.** These are classed as straight or flat chisels with a broad edge (Fig. 16, *a*), cross cut chisels with a narrow edge (Fig. 16, *b*), half-round chisels (Fig. 16, *c*), and gouges (Fig. 16, *d*). Chisels are made of steel with hard tempered cutting edges, but not hardened heads, and are from 3 to 8 inches long.

The **Cutting-chisel** is wedge-shaped and has a shank, which is set in the anvil under the object to be cut (Fig. 16, *e*). The cutting can be done on both sides simultaneously, a cold or hot set being used.

Small and thin iron, and wire are cut with,

Cutting-plyers or **nippers,** the blades of which are of sharpened steel.

WORKING TOOLS AND MANIPULATION. 29

Shears are used to cut sheet metal, hoop and flat iron, as well as wire: There are lever shears, the cutter of which works on a pin in a fixed lower jaw; parallel shears which work parallel with a drawing motion, and circular cutting pliers or nippers, the

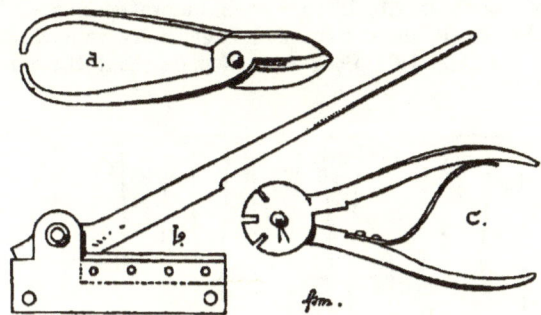

Fig. 17. Shears and cutting-pliers or nippers.

jaws of which are round, work on a centre-pivot and overlap but moderately.

Hand-shears, serving to cut off small and thin pieces of metal, are similar to ordinary scissors, only proportionately stronger in the

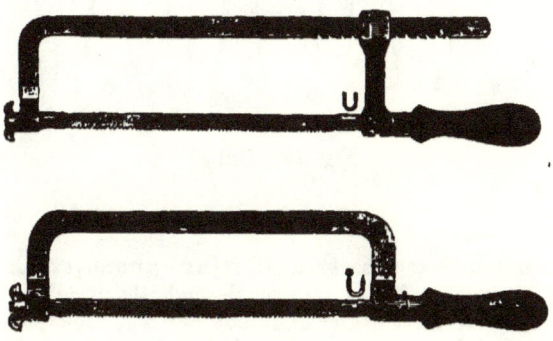

Fig. 18. Metal saws.

front part; the hinder parts being like those of the flat-nosed pliers (Fig. 17, *a*).

Stock-shears are used to cut heavier metal and have a fixed jaw, over which a one-armed lever works on a pin like a hinge (Fig. 17, *b*).

Fig. 17, c, shows a pair of **Wire-pliers,** the action of which is made clear by the drawing.

Circular and **parallel shears** are mostly powerful tools and are also much used in machine-work, serving to cut strong sheets, &c.

Saws are little used in art smithing. Metal saws have a more or less hollow bow, or rib, like the well known cock-saw. The blade is narrow, has small, unset teeth and is generally thinner at the back than at the cut (Fig. 18). Cock-saw work is about the same in metal as in wood.

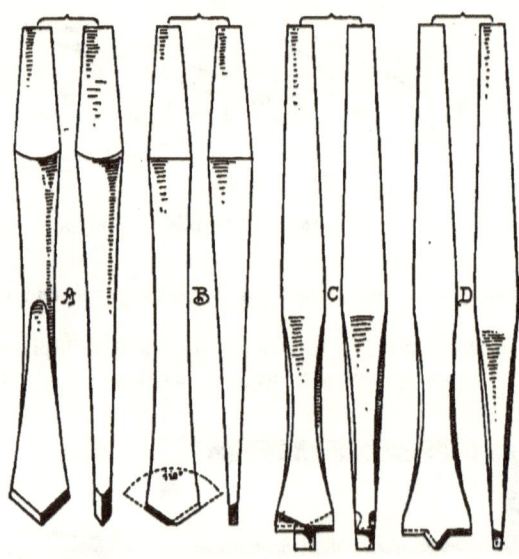

Fig. 19. Drills.

e. Punches and drills.

There is a difference between drifting, punching and boring. In the first a hole is driven into or through the metal without loss of weight. In the last the metal is cut out and removed.

The **Bolt-chisel** is generally a half-round chisel which is driven with the hammer. The enlarging and correct formation of the hole is continued by means of

Steel punches, round square or rectangular, &c. in shape and of various sizes. The lower end is set in the anvil; these tools taper towards the point. The hole is driven first from one side and finished from the other.

The **Punch** (used with or without handle) serves to make holes.

Under the iron that is to be pierced is placed a hollow iron cylinder, the opening of which is somewhat larger than the diameter of the intended hole. The enlargement of the hole may also be effected with the punch.

Thin sheet- and hoop-iron may also be pierced by means of a **Hollow-punch,** lead or wood being put underneath. The hollow in the punch is circular. The discs cut out gradually rise in the shaft as fresh holes are cut and fall out through the top end.

Punching-machines are often combined with shears. The punch, according to the nature of the machine, is worked up and down by a screw and centrifugal balls or with hand lever or by some similar contrivance. Such machines are also used for pressing and stamping sheets. Whereas in drifting (except with the hollow punch) no metal is lost, while in punching the piece comes out entire; in drilling, which is done by a rotating and pressing motion, the material comes away continuously in the form of small chips or powder, the drill falling through the hole. The motion is generally imparted to the tool, very rarely to the article being drilled. With light and simple machines the motive power is the hand; heavier machinery may be run by hand, by foot or by mechanical power (steam, &c.).

Drills are made of steel and hardened to yellow temper. At the shoulder they are tri- or quadrangular, or pyramidal or conical and tapered, this part being fixed into the tool or machine. The different forms have distinguishing names. The best known and most used are:

the **Lip drill** (Fig. 19, A), which cuts either to right or to left; it only serves for making small holes and is an imperfect tool;

the **Flat drill** (Fig. 19, B), cutting in one direction only and making holes up to about $5/8$ths of an inch in diameter;

the **Centre-bit** (Fig. 19, C and D), with a centre point which governs the motion;

the **Half-round Bit** (Fig. 20), used at the turning-lathe and producing a smooth wall or side;

the **Twist drill** (Fig. 21), which is the best, most rational and throws the chips well off.

Of the number of contrivances used to put drills in motion may be mentioned:

the **Fiddle drill,** used for very small holes. The gut-string of the bow is twisted round the roller of the drill, the rotary motion is imparted by drawing the bow backwards and forwards as in playing the fiddle;

the **Archimedean drill,** also for small holes only. The motion si obtained by running a loose socket up and down a spiral shaft;

the **Hand-brace,** so called because the necessary pressure is derived from the weight of the chest against the knob;

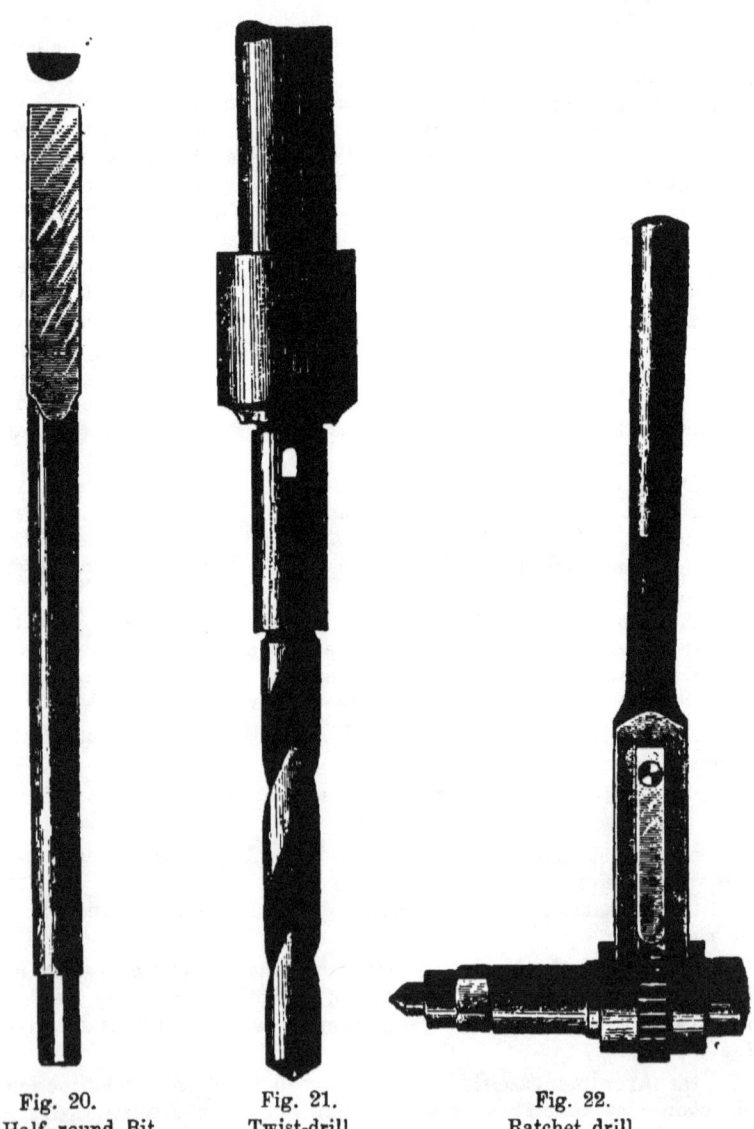

Fig. 20. Half-round Bit. Fig. 21. Twist-drill. Fig. 22. Ratchet drill.

WORKING TOOLS AND MANIPULATION.   33

the **Angular brace** for boring in a corner is in various forms. The motion is not direct, but transmitted;

the **Ratchet drill,** named in Germany from a rattling noise when in use. Fig. 22 shows one of its many forms. The drilling is effected by intermittent motion, the ratchet wheel or pinion checked by a spring detent, serving to prevent a backward motion of the lever.

Drilling-machines are constructed in so many forms that a full description of them here is impossible. These machines are classed as hand or automatic; and are further distinguished as independent, fixed, portable, &c.

In boring holes of large diameter the cut is annular and leaves a round core in the centre. Sunk holes, that is holes with the widest part at the bottom, are bored with unsymmetrical drills. Soft cast-iron and brass are drilled dry; malleable-iron and steel require lubrication with oil or soap and water.

Roughly punched or drilled holes are smoothed with

the **Rimer or Broach**. These are slightly tapering borers of various sections with handles, by which the defective hole is smoothed (Fig. 23, *a*). The best cross sections are those shown in Fig. 23, *b* to *d*.

The **Counter-sinks** may be mentioned in conclusion as tools serving to produce the holes intended to receive flush-screw-heads. Fig. 23, *e*, shows a rose-bit for conical, and Fig. 23, *f*, a sinker for cylindrical screw-heads; instead of the first named ordinary drills of larger size are sometimes used.

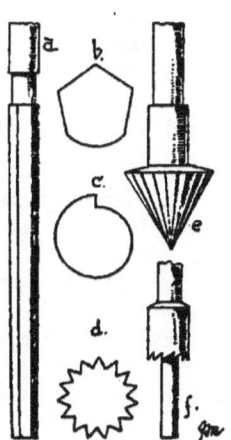

Fig. 23. Broaches, rose-bit, chamfering-auger or sinker.

f. Screw-making apparatus.

Inasmuch as screws play only a subordinate part in art smithing a few remarks concerning them must suffice.

Two parts are requisite in screwing, namely: the screw, a bolt with a spiral thread, and an internal screw or hole (generally some part of the object under manipulation) with a corresponding spiral groove. There are right- and left-hand threads; the former being in most general use.

In both the internal and external screw the thread and grooves succeed each other regularly. When the threads and grooves are triangular the screw is described as sharp cut; when the section is square it is termed a flat or square-cut screw. Sharp-cut screws are generally used, especially in small sizes and where they are required to secure objects firmly. Flat-cut screws are suited to large dimensions, more particularly where motion is to be produced. Leading screws are sometimes dual or compound, the spiral thread being once or twice grooved. Fastening screws are always simple. Whereas **wood screws** taper (screws for fixing wood), **metal screws** (for fixing metals) in addition to other peculiarities are always cylindrical. The diameter of a screw is that of the thread; the diameter of the nternal screw is that of the channel or groove. The various descriptions of sharp-cut screws are determined by the diameters and the felative acuteness of the above mentioned triangular thread. The English or Whitworth screw is the most used; an angle of 55° forms its basis.

The **screw-stock-and-die** is used in making screws. The thread of the internal screw in the die is cut by means of a

Screw-tap. These are separated into the taper tap, second tap and plug tap, or at least into the first and last. Small screw-taps are worked with the hand vice, larger sizes with special **tap-wrenches.**

The threads of external screws are cut with

the **Screw-plate** or **die,** a hardened steel plate with tapped screw-holes, this tool only serves for small sizes. Larger ones are made with

the **Screw-stock,** the forms and make of which are numerous. They have generally one feature in common, namely, that two, three or more taps with the cutting mother-thread are set in a frame which is screwed up from both ends. There are hinged, oblique, Whitworth and other screw-stocks.

The lathe is also used in making all varieties of screws, the cutting-tools consisting of an inside and an outside cutting-steel. Screw-cutting machines are also in use.

g. Appliances and tools for working cold surfaces.

The most important tools for finishing-off work are the **files.** These are made of hardened steel and are only smooth where they fit into the handle. Good files are pale grey in colour. They are first cut by the file-cutter in fine grooves or teeth set forward and then hardened. Single-cut files are grooved, as they have parallel cuts in one direction only, while double-cut files are teethed inasmuch as the first cut is followed by a second diagonal and somewhat narrower upper-cut. The number of cuts to the inch, giving the degree of fineness or coarseness to the file, is determined by the size and the purposes for which it is intended. These tools are known as rough, middle, bastard and second cut, and the subdivisions of smooth-files (smooth, dead smooth, extra smooth).

The forms of files are also numerous and vary according to their respective uses. The most common forms are:

Flat files rectangular in section, somewhat bulged in the direction of their length, tapering, and cut on three sides. If of equal breadth throughout they are called equalling and parallel files; if tapering towards the point they are termed cotter files.

Triangular files are equal-sided in cross section and taper to a point; the are known as 3-square taper, and Saw files.

Square files are four-sided in cross section, bulged and tapering.

Knife files, look like a coarse knife blade and are trapeziform in cross section.

Round files are circular in cross section; they are bulged, tapering, and have mostly a single-cut (rat-tail files).

Half-round files are semi-circular or show a smaller part of a circle in cross section, tapering, with a single-cut on the round and a double-cut on the flat side.

Entering or cross files are elliptical in cross section.

The object to be filed is generally held in a vice, and the file pressed in a forward direction, a backward or a double stroke being rarely resorted to. The coarse files are first used and afterwards the finer sorts, the finishing work being sometimes lubricated with oil.

Of machines used in working and finishing surfaces, the most important is

the **Lathe.** This is to be found in most lock-smith's workshops. It serves not only for turning round objects and surfacing, but also for various other kinds of work, as screwing, countersinking, spinning, drilling, grinding and polishing. The motion is obtained by means of a treadle worked by the foot or by mechanical force. The forms and construction of a lathe are various. Some general points must be noted. On the left side of the frame or bed is the fixed headstock set in motion by a small wheel connected with a larger disc below (serving as fly-wheel), and to the lower frame by a strap or cord. On the right is found the sliding-puppet or loose headstock with back centre which can be moved on the slide or cheek of the frame. The work is fixed between the fixed and loose headstocks, between which is the hand rest, serving to steady the hand and the turning-tool. When the work is guided not by the hand but by mechanical means the sliding rest replaces the hand rest. If the shifting of the support is effected by hand the engine is called a hand-rest-lathe. If the shifting is self-acting by means of a rack (connected with the driving-gear through a conducting-spindle) the machine is called a self-acting-slide-lathe.

Lathe tools serve for cutting. The hand tools have wooden handles; the slide rest tools are arranged for setting (fixing). For both uses there are **Gouge-tools** with curved cutting edge for the preliminary work, **Point-tools** of square steel cut diagonally with oblique edges running to a point, and **Flat-tools** with straight, chisel-like edges for finishing work, and, besides these there are cutting and **Inside-tools** for hollow turning.

Planing machines and **Shaping machines** are not indispensable in ordinary smiths' and skilled lock-work, so that only a brief reference is made to them here. In the former the work is fixed to a table which moves backwards and forwards in grooves on a bed beneath the fixed cutting-tool, the cut being one way only. The motions by which the object is gradually cut smooth, &c. are regular and automatic.

The shaping-tools are edged in various forms, set in revolution by the machine and serve to cut grooves, beadings, &c.

2. THE MANIPULATION AND TREATMENT OF WROUGHT IRON.

The processes used in wrought iron-work, not described in the preceding chapter, such as drilling, punching, swaging, &c. are briefly treated in this.

The **Forging** of iron on the anvil with the hammer is most satisfactory when the metal is heated to a bright red. Small pieces are forged by a single workman; larger ones require one or more strikers. The red-hot iron is cleansed from cinder by knocking it on the anvil and giving it a few light taps with the hammer; the neglect of this precaution may cause the cinder to be worked into the metal. If the latter is to become hard and elastic the forging continues until the iron cools, or else both hammer and anvil are wetted. Wet forging also gives a smooth surface. Objects that have become too hard are annealed, *i. e.* reheated to a faint red and allowed to cool gradually.

Welding, *i. e.* forging two separate pieces into one, is done under strong white heat. Both pieces must be equally hot. As the parts to be welded must be as clean as possible they are sprinkled with loam, arenaceous quartz, borax, sal-ammoniac or some other substance which prevents oxidation. The striking should be quick light at first and gradually heavier, and proceed from the middle outwards so that the slag may be thoroughly driven out and uneven spots be prevented. Welding by overlapping naturally forms a better junction than butt welding when the ends are only brought together: this is a reason why the parts should be prepared as above described, spread out, thinned, &c., prior to the actual joining. The welding together of iron and steel (which takes place principally in tool making) is called steeling, and requires great care and specia welding powder.

Flattening or taking down is equally necessary for lengthening or widening an object. The blows given with the narrow edge of the hammer are more effective than those with the broad face. The grooves wrought by the former are smoothed out by the latter.

Upsetting, or **jumping** is the exact reverse of flattening; it means both thickening and shortening. It is effected by striking the red-hot piece on the anvil or on a special jumping-block, or by hard-hammering on the end.

Bent or twisted pieces are **straightened** to the right shape. It is generally done with the hammer on the anvil, but sometimes a special straightening-plate is used; the process may be effected in either hot or cold state. Flattening, jumping and straightening require a certain dexterity in handling which is not easy to describe.

Bending may also be done warm or cold. Bending at right-angles is done by hammering over an edge of the anvil, or the square part of its beak, or with the aid of a vice. Curves are produced on the round end of the beak or on a conical mandrel. In curved bends a scroll-horn is also occasionally used. This is a tool which is fixed into the vice

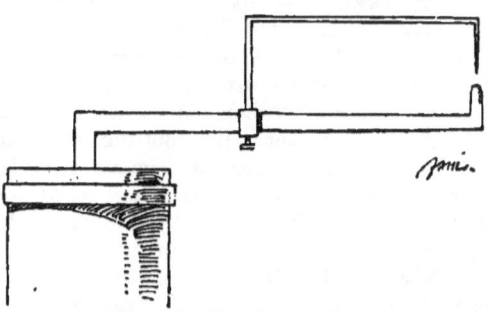

Fig. 24. Snarling tool.

and has two cylindrical prongs forming a fork. For spiral windings and other forms which occur frequently in art smithing, special scroll tools, pins, &c., round which the object is twisted, are called into use. These tools have the same form that the piece being handled is to take; the latter is generally of thickish flat iron. Large flat curves are produced by placing the iron on two raised and separated supports and striking downwards in the middle. Sheet iron is best bent and turned over on bending and tilting machines.

By **Embossing** is understood the punching out of rounded bumps called "bosses". The iron may be bossed out when red hot by driving it into a suitable cavity or swage. Bosses are produced by hammering the iron when cold, with suitable ball hammers, upon an under-layer of wood or lead.

Large flat domes are obtained by hammering out the sheet-iron gradually and from the middle towards the edge. In finer work small bosses are produced by means of a special instrument, called a snarling tool, shown in Fig. 24 and is made fast in the vice or onto a special block. By striking with the hammer near the fixed end of the tool the arm vibrates and its bent striking end produces by percussion the bosses in the metal exposed to its blows. The upper part serves as a gauge for ensuring that the bosses are formed in the right place.

Impressing sheet iron into hollow and rounded forms is effected with the lathe by tool pressure.

Punching is done with tools of like name, which are short steel bars rectangular in section, with the edges taken off and tapering towards the point, which is of many shapes. Small bosses can be made with them and the punching-hammer in sheet-iron, as well as bead-like or ribbed surfaces. Whereas thin sheets are generally punched from the back so that the bosses appear in relief, with thicker sheets the ornament is indented with punches or chisels.

Engraving consists in cutting flat, mostly linear, designs in the surface of the sheet by means of the graver, or graving-tool. This is generally done by pressure of the hand; more rarely (when the work is heavier) with the aid of a hammer. When the latter is used and especially when curves are made it is called

Cutting in iron. Iron cutting, engraving and **chasing**, that is the finishing off of cast or hammered parts either with the graver, or with punches or other tools, are usually only used in iron in small artistic objects. These constitute an art in themselves and are rarely performed by the skilled smith or even locksmith.

Etching is effected by means of acids. The sheet is first covered with a layer of protective wax, asphalt, or some suitable varnish; the parts to be etched are then deprived of the protective agent and the acid eats into them to the desired depth. Where the surface which is not to be etched is smaller than that which is to be affected the reverse operation sometimes occurs, *i. e.* the protecting material is painted or otherwise laid on in the necessary places only. When the acid has acted enough it is cleaned off with turpentine. Etching is principally used to ornament smaller artistic objects, such as arms, &c. Sometimes the etched parts are colour-varnished so as to give the effect of Niello, or Enamelling.

Niello-work means that the metal basis is engraved in the same manner as in copper-plate engraving, and that the parts cut away are filled with a substance compounded of sulphur, silver, copper and lead. In melting in the niello compound the metal work must not be made red hot or it would waste and become holed.

In **Inlaying** or hammering metal into metal, as gold and silver on iron, the parts to receive the inlay have a dove-tail cut (wider at the base than at the surface), which is produced with a chisel and into which the softer metals are hammered. A simpler and cheaper, but also less durable way, is to hatch, and to cut in lines with the graver, and hammer the precious metal on to the roughened foundation. The unembellished parts are afterwards made smooth.

Enamelling, i. e. the melting on of vitreous paste, is almost exclusively applied at present to cooking utensils, baths, advertisements, &c., and therefore to useful rather than artistic purposes*).

Completely smooth and bright surfaces are produced by planing, grinding and polishing.

Planing is done with the plane, which is made in various forms, but must always have a sharp, faultless edge. The elevations to be removed are most easily detected by moving the surface under manipulation backwards and forwards on a perfectly level, painted straightening-plate.

Grinding is effected either with hand-grindstones or with a wet or dry revolving grindstone. Discs of emery, pumice-stone, emery-paper and cloth, emery-powder or iron-scale are also used with oil, on wood, leather or lead.

Polishing gives the object that degree of smoothness which may be described as glint. Is is produced by continuous friction with fine powder that is taken up by soft leather or wool, moistened with spirit or oil. Lime, rotten-stone, putty powder and crocus, serve, among other things, as polishing powder. Burnishing tools and agates are also used; these also serving to press down inequalities in the metal. The shapes of the steels, which are set in wooden handles, depend on the work. Round or cylindrical bodies are best polished on the lathe.

As iron and steel easily rust under the influence of the air, especially of moisture and wet, the surface is protected sometimes by means of other metals, or by bronzing, blackening or tempering, or by coating with varnish or oil-colour. In every one of these processes a thorough cleansing from scale, &c. and a metallically pure surface must first be obtained if the result is to be satisfactory, and rust must not be allowed to set in beneath the covering material. This is done by **pickling** or removing the surface with diluted sulphuric acid; or by reheating and also by brushing and scraping.

A **coating** of lead, zinc, tin, copper, brass, nickel, silver or gold may be deposited either by "dry process", in which the articles are dipped red hot into the molten metal, or by "wet process", when they are dipped into liquids which contain the metals and chemicals in solution necessary to effect their union; or through precipitation by galvanic action; or by plating, when the covering-metal is pressed or rolled on to the iron in the form of thin sheets or plates. With regard to gilding it must be mentioned

---

*) The iron-works at Gaggenau in Baden have of late produced artistic enamelled iron ware.

that other processes are known, namely mercury gilding, in which gold in amalgam with quicksilver is laid on and the latter evaporated by heat; and leaf-gilding, in which the gold-leaf is pressed onto a prepared and roughened metal surface and polished with a steel burnisher; or merely attached by size in the ordinary way.

Browning or **bronzing** consists in creating an artificially oxidised surface, the oxide protecting the metal from further rusting. (It is specially applied to fowling-pieces, &c.)

Blackening consists of smoking the article over a fire of resinous wood and then brushing; or the articles are smeared with linseed oil and this is burnt off over a fire.

Varnishing with a transparent mixture of linseed oil and turpentine protects uncovered surfaces. Oiling, or smearing with tallow may be substituted*).

Varnishing with iron-, asphalt-, or tar-varnish, or coating with oil-colour is chiefly confined to coarser articles having less carefully worked surfaces and to objects that are exposed to the weather. Before the actual painting takes place a ground coat, consisting of lead paint or graphite is laid on. In former times decorative effects were sometimes produced by polychromatic treatment, and of late attempts are being made to revive the fashion. There is scarcely any objection to be raised to this from the point of view of style, provided always that it is governed by good taste.

Fig. 25. Specimens of weldings.

3. THE ORDINARY METHODS OF JOINING IRON-WORK.

Various methods are used to unite and fasten together separate pieces of iron; the most important of which to the smith must now be briefly enumerated.

The **Welding** together of separate parts is the most effective, and the means best suited to the smith's craft, although it is not

---

*) A preventative against rust, called "Mannocitin", made by the firm of Ed. Müller & Mann in Charlottenburg, has been much recommended of late.

always the easiest method. As this process has already been described, it may suffice here to remark that for railings, balustrades and similar artistic iron-work, **welding** is principally adopted where scrolls or other forms of ornament consist of two or more parts as shown in Fig. 25.

Brazing or **Hard-soldering** (in contradistinction to "soft-soldering" with tin) forms a junction which will bear a certain degree of hammering and bending: copper, brass, and, where in finer work the red or yellow colour would disturb the effect, silver, are used. The parts to be soldered must be metallically clean and free from oxide; these parts are packed in loam which is made more adhesive by the addition of horse-dung, and, as an adjunct to the solder, borax or powdered glass is used. Red heat is necessary to soldering. Small objects are soldered with the aid of the blow-pipe; larger ones require a charcoal or coke fire. The junction occurs when the solder begins to melt, denoted when the flame turns green. In order to keep the parts in position while being soldered, they are bound together with wire, or temporarily riveted together, and so on.

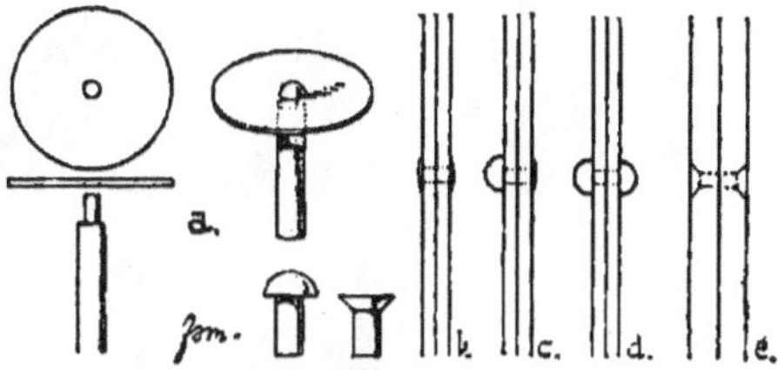

Fig. 26. Rivetings.

Puttying and **cementing** only secure a firm connexion when the individual parts already fit into each other, as with various sized tubings, and have of themselves a certain hold. Cementing is also employed to set iron in stone or other material.

Riveting is a method most frequently used. It will either render parts immoveable or act as a pivot on which they may turn. Either one part is tenoned to serve as rivet-pin while the other is drilled for the rivet-hole (Fig. 26, *a*), or which is the most common, both parts have holes through which the rivet is passed. The rivet is either a cylindrical pin which is hammered out broad at both ends (Fig. 26, *b*), or it has a head at one end, while the other end is hammered flat (Fig. 26, *c*), or else clinched into a shaped head with the riveting-set (Fig. 26, *d*), or lastly both heads may be sunk, in which case the rivet-holes are conically widened or countersunk (Fig. 26, *e*). Small objects are riveted cold, larger ones at red heat.

Screwing is more especially used in cases where it may be necessary to take the work to pieces again. Either one part holds the male- and the other the female-screw, or both parts may have an internal screw into which a separate external screw is driven. The screw may either have a cut-head, sunk or raised, and the tail end filed off flush with the surface; or it can have a head like a rivet, with a screw-nut under which a disc of tin is sometimes placed, fixing the other end. The first mentioned screws are made fast and loosened with the screw-driver, which resembles a chisel; whereas the nut is fixed and loosened by means of a wrench or spanner. This tool is made in specific sizes. There is also an adjustable screw-wrench which can be set to any desired size.

Riveting and screwing occur also, without mentioning countless other instances, when pieces of iron are drawn down and lapped or butted together.

Drawing down and **riveting** often replaces welding and is principally adopted in flat-iron scroll-work. It consists in placing

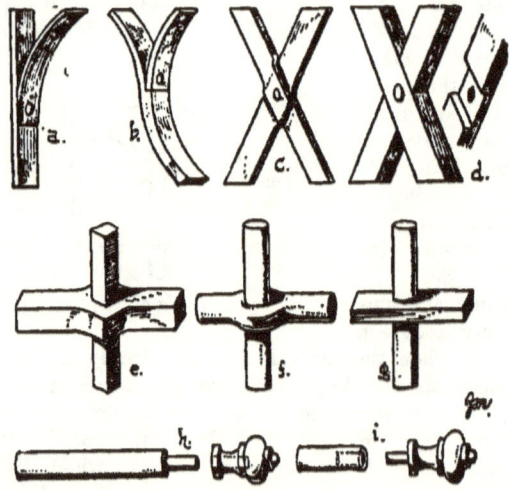

Fig. 27.
Various methods of fastening ironwork together.

a piece which is brought down to a thin edge against another (Fig. 27, *a*). If the piece that is to be so fixed is not thinned off, or only partially so, a step-like cut is made at the place where it is to join (Fig. 27, *b*).

Intersecting, especially when flat- or square-iron pieces cross each other, may be done without thinning either of the parts, which are bent outwards (Fig. 27, *c*), or each part is thinned or cut away to the extent of one half, so that they are flush on both sides (Fig. 27, *d*).

For **passing** one bar through another a fitting hole (Fig. 27, *e*, *f*, *g*) must be punched or drifted.

WORKING TOOLS AND MANIPULATION.   43

Tenoning and **pinning** are adopted more particularly to fix cast iron spear-points and pine-apples, &c., to balustrades, railings, &c. (Fig. 27, *h*, *i*).

A much used mode of fastening, where two or several parts are to be fixed together, is

the **Collar.** The iron used is generally rectangular or half-round in section (Fig. 28, *a*, *b*, *c*, *d*, *e*).

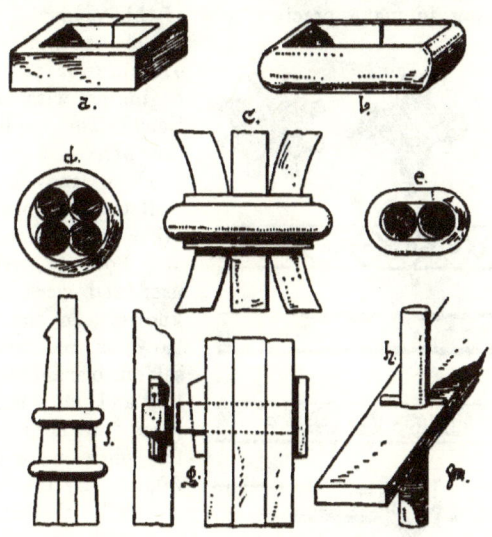

Fig. 28. Collars and wedges.

Sometimes the collars are tightened by means of a **Wedge** (Fig. 28, *f*). The wedge is an effective mode of joining and can be easily loosened again, but it is mostly used to finally tighten up work (Fig. 28, *g*, *h*).

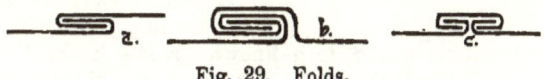

Fig. 29. Folds.

By **Shrinking on collars** is meant the hammering of red-hot rings, hoops, &c. over the parts to be secured. As they shrink in cooling they give a firm hold.

Folding is only used in sheet-iron. There are single- (Fig. 29, *a*) and double-folds (Fig. 29, *b*) and overlapping folds (Fig. 29, *c*).

4. WORK DETAILS OF MOST FREQUENT OCCURRENCE IN THE SMITHS' ART.

Although the number of these is almost countless and they vary materially in style with the various periods of art; some, which are constantly recurring and form, as it were, the ABC of the "form-language" of the smiths' art, may be mentioned. No pretence to completeness is made, yet such as are referred to may prove useful to those desiring to make practical use of this manual.

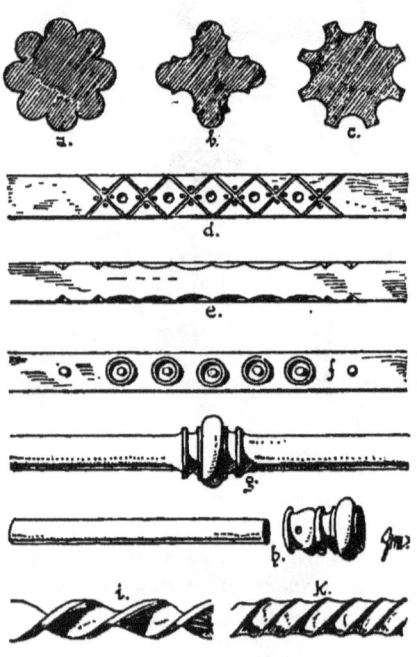

Fig. 30. Ornamented bars.

Let us first consider the ornamentation of bars, beginning with the cross sections. The rolling-mills of the present day are able to produce fancy bar-iron with stellate, cruciform, and many other such sections. These are, however, not frequently used and were formerly unknown. In the middle-ages bar- or rod-iron was not seldom ornamented by chiselling and punching simple patterns into it (Fig. 30, *d*), or the edges were fretted (Fig. 30, *e*). By the use of swages regularly shaped protuberances were produced (Fig. 30, *f*). Swages are also used to produce moulded swellings (Fig. 30, *g*). Of late this is done more simply, if less genuinely and solidly, by slipping malleable cast-iron sockets, &c. over the bar and riveting them (Fig. 30, *h*).

A good, effective and long-known process is the **Twisting** of bars while red hot, which can be done with the aid of the tongs in light work, but which in heavier work requires the help of a screw-stock or wrench (Fig. 30, *i*, *k*).

Scrolling into **volutes** is universally practised. There are many varieties of these. Thus the bar may be bent with the scroll wrench or round a scrolling iron without varying the section of the bar (Fig. 31, *a*), or it may be drawn down or snubbed, thus varying the cross section (Fig. 31, *b*), or it can be slit into 2 or 3 volutes

WORKING TOOLS AND MANIPULATION. 45

(Fig. 31, *c*). The inner end of the volute is often ornamented with a rosette or knop (Fig. 31, *d*).

The **Slitting** and opening out as a break in the length of a bar is effective but uncommon (Fig. 31, *e*).

Spindle-shaped spiral twists, especially of round rods and thick wires, are more common (Fig. 31, *f*).

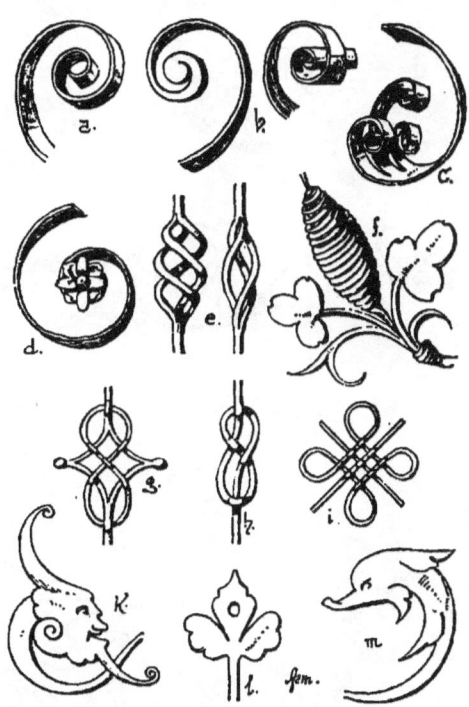

Fig. 31. Bars scrolled into volutes, slits, interlacings, spiral-twists; forged flat ornaments.

Repeated interlacings, a kind of plaiting, are favourite forms in the renaissance style (Fig. 31, *g*, *h*, *i*).

The **Hammering out of bars** into **Flat ornaments,** such as leaves, masks, or grotesques, also often recur in the same period of art (Fig. 31, *k*, *l*, *m*). The outline is cut out with the chisel or shears and then finished off with the file.

The **Beating of scroll ends** into **forged** or **embossed leaves** carries the art a step farther, and reached great technical perfection

46   SECTION II.

Fig. 32. Embossed leaves and acanthus husks.

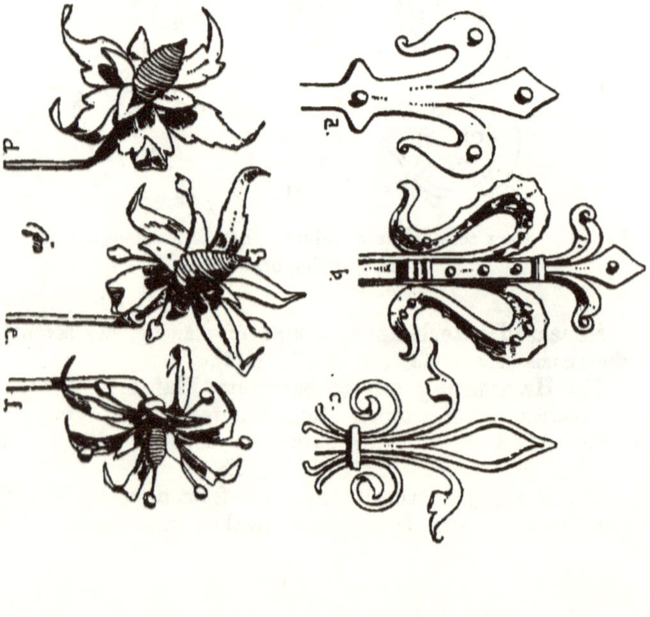

Fig. 33. Examples of lilies and other flowers.

WORKING TOOLS AND MANIPULATION. 47

Fig. 34. Examples of rosettes.

Fig. 35. Spear-heads and knops.

48   SECTION II.

in the baroque and rococo ages. The principal natural motive is the acanthus. Fig. 32 shows a number of foliage patterns (*a* to *f*).

Fig. 36. Cartouches, shields and masks.

The **Acanthus husks** may be mentioned here. They serve as ornamental envelopes to the bars, or as free ends, or cappings to them (Fig 32, *g, h, i*). One often finds in the latter position

Fig. 37. Wreaths and sprays.

Lilies (fleur-de-lis); these were often used as motives in the middle-ages and frequently recur later (Fig. 33, *a, b, c*). They are only one of the peculiar forms of

WORKING TOOLS AND MANIPULATION.

the **Flowers** used in smithing of which others are given in Fig. 33, *d*, *c*, *f*.

Rosettes are to a certain extent fanciful renderings of flowers. Fig. 34 shows simple and richer examples. Whereas formerly these were all forged in swages, or embossed by hand, they are now often stamped out by machinery. The latter work is more uniform, and also more monotonous.

Fig. 38. Specimen of Applied work.

The like may be said of **Spear-heads** and **Knops,** which serve to finish off the tops of railing bars and the like. These heads were formerly forged by hand, while at the present time they are largely stamped out of red hot iron or produced in malleable or common cast-iron. Examples are shown in Fig. 35.

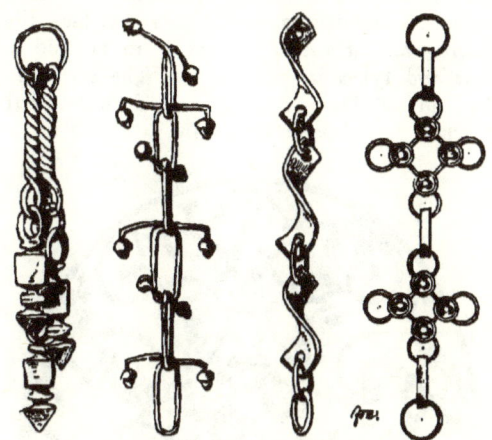

Fig. 39. Ornamental chains.

Cartouches and **Shields** cut out of sheets, bent, embossed, and ornamented with scrolled volutes, are often used as decorations to grilles (see Fig. 36, *a*, *b* and *c*).

Lockplates, Escutcheons and similar objects were in the middle-ages, the renaissance, baroque, and rococo periods mostly pierced into open work, slightly chased, embossed, &c. Fig. 36, *f* shows an example from Wertheim on the Maine.

Masks and **grotesques**, occur not infrequently in richly forged work, and are mostly embossed out of sheets, less frequently forged from the solid (see Fig. 36, *d, e*). Such details require a very skilful hand and artistic capacity, if they are to prove satisfactory, otherwise they are best left alone. Herein the limit, of that which appeared to be feasible, both in respect of material and of technic, has been reached, — if not already overstepped.

Garlands, festoons, wreaths and **sprays** representing natural flowers also require skill and taste. They are in other respects easy to make and rarely fail in producing a good effect. They appear in grilles as subsidiary and ornamental embellishments, and on wrought-iron works of larger dimensions. Fig. 37 shows three specimens coming under this heading.

By **Applied work** is understood the process whereby sectioned bars and ornamental details are applied to smooth iron surfaces or stout iron sheets as shown in Fig. 38. Such applications are most frequent in locks and their mountings.

Lastly, reference must be made to **Ornamental chain work,** which may take very many forms and styles, according to the purpose for which it is intended, and to the dimensions required. It is used for hanging coronas, chandeliers, wall-lights, &c. (see Fig. 39).

The foregoing almost exhausts what is to be said concerning the individual forms and types in general use. The sections which follow will, with the aid of their illustrations, furnish further data in elucidation of what has already been said.

SECTION III.
HISTORICAL DEVELOPMENT OF THE ART OF SMITHING.

1. THE ANTIQUE.

Although it will be endeavoured in this section to give a picture of the historical development of the technics of smithing, no attempt will be made to deal with the question of how iron was produced in earlier times, about which only incomplete information has come down to us, and especially as archæologists and experts are in some respects in disagreement regarding it. Those, however, who are interested in this part of the subject are referred to the highly meritorious work written by Dr. Ludwig Beck on the History of Iron in its relation to technical and historical culture development, published by Vieweg & Son, Brunswick.

It is now known that the production of iron and its use are of very ancient date, far older than has usually been supposed and certainly dating back to prehistoric times. In the British Museum a piece of iron is to be seen which an Englishman, J. R. Hill, found in one of the inner masonry joints of the great pyramid of Cheops. This broken part of a working tool probably shows the greatest, historically proved antiquity, namely about 4000 years. Articles of iron found in other places and the wall-paintings on their graves show that the ancient Egyptians used iron weapons, sickles and other tools, ship-sheating, &c., which they must either have made themselves, as was probably the case in most instances, or which they drew from Ethiopia, the inhabitants of which pursue the iron industry to this day.

Iron was equally known in ancient Assyria and Babylon. The excavations have brought to light among other things iron finger-rings, bracelets, weapons, chains, hammers, knives and saws. Victor Place even found a complete iron store at Khorsabad. The principal part, estimated at 358 tons, consisted of pieces of iron pointed towards each end and having a hole near the one end, which were identified as unwrought ingots. The perforation was probably to facilitate transport, by stringing the blocks together.

In Phoenicia and Palestine iron also came early into use. In the Bible (Genesis Ch. 4, v. 22) one reads that Tubal-cain, the son of Lamech and Zillah, was "an instructor of every artificer in brass and iron". The like was the case in Persia, India, China and Japan. The Chinese claim that steel was invented 2000 years B. C., and Indian steel was also favourably known long before our chronology. Further evidence of the antiquity of iron-working is found in philological comparisons. The Sanscrit word for iron is "ayas", the Persian (Zend) "ayanh", Old Gothic "ais", Old High German "aisin", "isan", "isen", Anglo-Saxon "iren", English "iron", Old Norse "iarn", Swedish "jarn", Spanish "hierro", Italian "ferro", Latin "ferrum" and French "fer" (see Beck's work).

Culture came from the East, from Egypt and from Western Asia to Greece and thence to Italy, and after what has been said above, it would be too strange to even suggest that the Greeks and Romans did not know and utilise iron. That they did know it and understand how to work it, is shown in many of their writings; it is made evident by their painted vases and bas-reliefs and it is proved by the (few) iron articles which have been discovered.

Iron and steel were already known to Homer. Schliemann disinterred iron objects in Troy and Mycena. Glaucos of Chios (600 years B. C.) is held to be the inventor of welding or soldering iron. Not only were weapons of attack and defence, agricultural implements and all sorts of objects made of iron; the metal was used also for ornamental vessels and statues, the latter being made of embossed pieces which were afterwards put together. Thus, we hear of an artistically wrought-iron base to a silver vessel at Delphi, of an iron statue of Hercules, &c. Various Greek cities, as Corinth and Athens, had recognised markets for ironwares. Although the best steel came from Chalybes and India, the Laconian and Lydian steels were also prized. Smiths' tools, as they are pictured on Grecian vases, represent anvils, hammers, pincers, &c., and even their bellows are to all intents and purposes the same as those in use at the present day.

The articles found in Etruscan and Roman graves, the excavations at Pompei, Vulci, Cervetri, Caere and many other places have also brought iron weapons and utensils to light. Searing-irons, fire-hooks, tripods, locks, keys, brasiers, money-chests were often made of iron;

so likewise were the weapons for use, whereas those for ornament were made of bronze or brass. It was customary to wear iron rings as the insignia of a free-man and, probably, also for use in sealing up doors, &c.

Even if in the early ages iron was principally imported in Italy from the Island of Elba, the Romans, on the other hand, after having acquired the sovereignty of the world, most undoubtedly produced and worked iron in various provinces, for instance, in Spain, on the Rhine, in Carinthia, and it may well be assumed that they found this industry already in existence in such parts.

When we attempt, and it is almost in vain to do so, to sum up the position of the smith's art in antiquity, the following points are salient. First, iron is very much more sensitive to oxidation and to rust, than bronze. The most of that found has actually turned to powder and dust, and that which still exists is eaten away and unattractive. There can be no doubt that the ancients, as the rule and not the exception, only used iron and steel where probably no other material would so well answer the purpose, such as for tools and weapons, and these of the simplest forms that would serve the purpose. For show and for articles of luxury the brilliancy of bronze and of the precious metals was given the preference. Ordinary labour was performed by the slave; skilled labour could be undertaken by the freeman: We thus find a second reason for assuming that things which were to show artistic finish, such as bronze lamps and vessels, were hardly ever made of iron. In any case, that which modern museums have to show of antique iron-work cannot be for one moment held in comparison with the bronze- and brass-work, the ceramic and similar arts of the same period.

The state of the ancient iron industry may be briefly summed up. Greeks and Romans knew iron; they produced it in open hearths or in small ovens with the aid of natural wind draught or by bellows; they thus produced a material bearing sometimes the character of malleable-iron, and sometimes that of steel; they usually employed it for articles which could not well be made of other material and only very exceptionally gave these objects a distinctly artistic form. To cast-iron and to the production of malleable-iron and of steel as known to the present age they were strangers, and, in consequence of the then state of science and of its technical appliances had naturally to remain so. Antiquity has thus but slightly affected the later development of the smith's-art and its influence is therefore imperceptible.

2. THE MIDDLE-AGES.

During the collapse of the universal sovereignty of Rome and in the confusion arising from the migration of races a great portion of antique culture became lost and therewith much of the highly developed technic both of art and of manufacture. But this statement can scarcely be said to apply to the art of smithing. First of all, as seen in the last chapter, the iron industries of the ancients were, in respect of art, of an entirely subordinate nature, and, on the other side, the unending campaigns and wars, which arose in the stage between the old and new periods of culture, were enough to secure the progress of at least one branch of smithing, namely the armourers, which necessity compelled for good or evil to stride in the direction of development. Even admitting that the requirement of such times made the quality of the material used, and its practical fitness to the aim in view, rather than mere external forms, the chief question, yet nevertheless the seeds of further development were sown, even of the very forms used in later and more peaceful times.

The middle-ages thus had the privilege of introducing smiths' work into the regions of architecture and to some domestic uses, and of discovering styles suitable to these requirements. In this mediæval nations were successful in a high degree. Artistic smithing of the middle-ages has indeed handed down to us specimens which show astonishing skill and a fine sense of form. But our wonder is increased when we realise the exceeding simplicity of the appliances wherewith such results were brought about, and when we reflect that hammer and anvil were, as a rule, the only tools used, and that each rod, each wire, each sheet had to be wrought, and that neither these nor the rolled material obtainable now in every form and size, were to be had ready-made when they were produced.

It must, however, be remarked that perfecting the technical means of production does not invariably tend to render the artistic wares of handicraftsmen more perfect. Closer consideration makes this truth more apparent. It must, among other analogous examples, be obvious that repeated forging and welding improves the quality of iron; but, not only did manual labour furnish a better iron than that averaged by the mechanical operations of the present age; but the external appearance of handwork has something fresher, more original and interesting than machine productions, although the latter are undeniably neater and more exact in appearance. Let hand-needlework be compared with machine embroidering, and the same conclusion follows, and this applies to other branches of art with equal force.

But then, hand work with the comparatively simple tools employed demanded also a great sacrifice of time. Machinery, as the substitute for handicraft, owes its existence to the striving after shortening and lightening of labour and to the ensuring cheapness of production. For these, as well as for other technical reasons, the older manual work was not in a position to produce objects of large dimensions;

when these were here and there successfully accomplished the results are surprising and entitled to the highest appreciation.

Putting the armourer's art aside for the present, the fourth section of this manual being specially devoted to it, the smiths' work of the middle-ages in connection with architecture and industry commenced from about the 10th century to deserve notice. At least it is to this period that the oldest work which has come down to us dates back. In the 12th and 13th centuries the smiths' productions become already highly important.

Here too it was that the church enlisted art into her service and gave the most important orders. Let us first take note of the ornaments for doors and gateways, and for chests and presses, of the window-fastenings and grilles, of standard and hanging candelabra

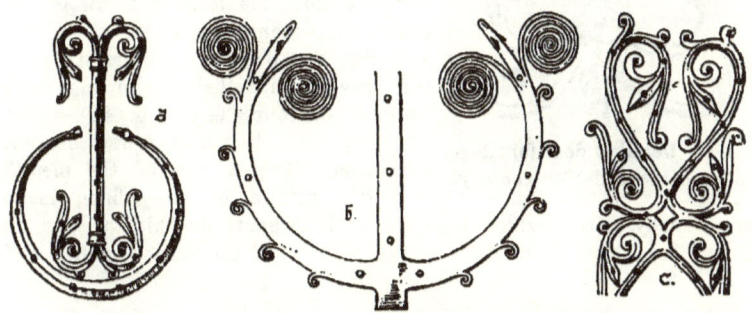

Fig. 40. Parts of Romanesque door-furniture.
a. The Cathedral of Puy en Velay by Ebreuil. b. The church of Blacincourt, Gironde. c. The church of the holy sepulchre, Neuvy. 12th century. From Viollet-le-Duc.

For secular purposes, such as the embellishment of castles and of corporation buildings, we find other important items including firedogs and other hearth-furniture, wall-anchors, door-knockers, &c.

The external appearance of the smiths' work of the Romanesque age presents little that is elegant; the forms are full, heavy, and give the impression of great solidity. They accord in their simplicity with the styles of architecture and ornamentation prevalent at the same period, and present similar characteristics. The most elegant and the richest work is found in door-furniture, belonging especially to the latter part of the Romanesque age and in the period of transition to Gothic. It corresponds with the wooden construction of the middle-ages with its small, grooved and tongued narrow boards, which, in themselves, afforded little scope for ornamentation until ironwork began to be spread over the large flat surfaces. Although the original purpose may only have been to join the wood-work well and securely, the spirit of decoration soon assumed the foremost position. Instead of the simple tongue, angle, and cross garnet hinges, and the crescent-shaped straps, which were especially favoured in the earliest part of this period, rich scrolls, twining over the whole of the door and forming a peculiar ornamentation, began to appear. Noteworthy door-furniture of this kind is found in the cathedrals of Paris, Liège, and Rouen, all of which date from the 13th century.

Fig. 41. Detail of door-furniture, Liège cathedral. 13th century.

Characteristic features of Romanesque iron work are the slitting of bars and scrolling the parts (see Fig. 40, *b*), the welding together of separate bars into complex bars, the forging in swages of ornaments, such as rosettes, stars, &c., and also the peculiar conformation of the leaves, with their hollowings and rounded contours (see Fig. 41). This work was all, to use the now current expression, "forged out of the piece", *i. e.* it consisted of one whole, made up of many parts welded together, and without the aid of screws, rivets, &c. Most distinctive is the technic of the forged grilles and

Fig. 42. Details of a Romanesque fire-guard. 13th century.

of the implements of that period; which instead of being secured together with nails are bound with ties or collars (see Fig. 42).

HISTORICAL DEVELOPMENT OF THE ART OF SMITHING. 57

In the transition to Gothic the technics were changed and developed. Besides "forging out of the piece" and welding up, cold-

Fig. 43. Detail of Gothic grille from St. Denis. 14th century.

Fig. 44.
Gothic ornamental detail.

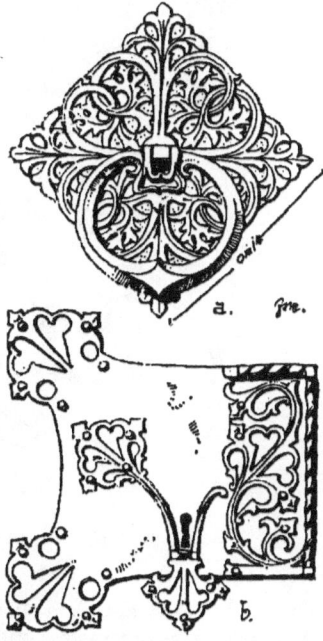

Fig. 45. a. Door-handle of St. Peter's, Strasbourg, 15th century, backed with red cloth.
b. Gothic ornaments from Münster in Westphalia.

riveting was also employed; individual swages, or loose forgings were riveted onto the principal parts (see Fig. 43).

The section of the leaf-work became altered; the bars being hammered out to thin sheet-like metal and cut into definite outlines, as well as bent, stamped out, or embossed (see Fig. 44). Bars were also twisted. Graving-tools, chisels and punches are added to the

Fig. 46. Gothic details in wrought-iron.

tools in use. The complete effect becomes richer and more animated. This improvement increases gradually till Gothic art reaches its zenith. Boldly curved, long drawn out designs, crab like leaf forms characterise this period, in which already all sorts of articles were made of iron; elegant chandeliers and lanterns and even iron-furniture. Much flat ornament was required for lockwork. Keys were also sometimes ornamented. The embellishments were tastefully fretted and

HISTORICAL DEVELOPMENT OF THE ART OF SMITHING. 59

their effects enhanced by a backing of coloured cloth or leather, &c. (see Fig. 45). Scarcely any other material was so well adapted as

Fig. 47.
Late Gothic door-knocker. 15th century. In private ownership at Augsburg.

Fig. 48. a. Details of the well cover near the Cathedral at Antwerp. b. From the Cathedral at Prague. 14th century. c. Late Gothic door ring.

wrought-iron to the principles of decoration used in the nobler periods of Gothic art (see Fig. 46).

The later degenerate Gothic style created much that is not decidedly tasteful or consistent, inasmuch as it shows the stilted and heavy work, introducing the fish-bladder and other unsuitable motives to the graceful styles of wrought-iron work (see Fig. 47). It was at this period also that a questionable naturalism was introduced, in the shape of gnarled branches serving as door-knockers, &c. (see Fig. 48).

The production of stone and wooden profiles in iron will be discussed in the next chapter. If we summarise the results of the development of wrought-iron work during the middle-ages, it will be seen that, with comparatively simple appliances, work on an important scale was executed; that in respect of technical routine, and richness and variety of artistic effect, it did not equal later styles; though on the other hand, it carried the constructive principles of wrought-iron to a degree of perfection that has hardly been equalled since. It is, moreover, clear that the middle-ages are entitled to the credit of having made the first attempts to treat wrought-iron polychromatically, to give it the effect of colour by means of paint and at the same time by such means to afford it protection from the destructive influence of rust.

An exhaustive and richly illustrated description of the smiths' work of the middle-ages is given by Viollet-le-Duc in his "Dictionnaire raisonné de l'architecture", volume 8, under the heading of "Serrurerie" (Locksmith's work), to which work special attention is called.

3. THE RENAISSANCE PERIOD.

It follows as a matter of course that as the mental and cultured life of a people rises or sinks, so the arts advance or recede in sympathy. This is seen in the transition from the antique culture to that of the middle-ages; and even more in the mighty progressive change from the latter to that of the renaissance period. The struggle for mental freedom, the striving to substitute a principle allowing of life-like and fancy-free action for one confined within a strict, dry and narrow set of rules, finds most vivid illustration in the emancipation of the arts during the renaissance. If this was not equally the case in each particular branch of art, this arose from the fact that many varying extraneous circumstances influenced the process of transition in greater, or less, degrees. We have to consider carefully whether the technical results in any particular field of art reached their highest point, or only a stage in the process of development, during the transition. Further it is a long recognised fact that the minor arts are dependent on their mother and teacher, Architecture, for their growth, just as children are dependent on their parents. The minor arts, speaking generally, require a generation for the influence exercised upon them by architectural changes to be felt to the full extent. The traditions of handicrafts are more unyielding, than those of high art.

On these and other grounds changes in style are not accomplished suddenly and violently, but by degrees; a mixture is found, an intercalation, or amalgamation of the preceding with the succeeding styles. This amalgamation brings together things of doubtful style with those which are naive and charming in the highest degree, and to which a certain originality cannot be denied. We see this particularly in the ornament of the transition from Romanesque to Gothic; but it is still more pronounced in that between Gothic and renaissance, or, as it is called, "the early renaissance".

Breaking off these observations in order to revert to the object of this manual, it must be understood that, while in architecture and wall-painting, art reverted to antique models (hence the term "renaissance", meaning a new

Fig. 49. Details of a late-Gothic bracket. 15th century.

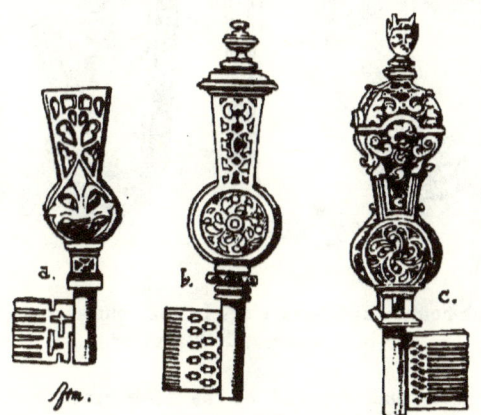

Fig. 50. Keys of the 15th, 16th and 17th centuries.

birth), such could not be the case with smithing, simply because this latter branch of art was, comparatively speaking, very little developed

SECTION III.

in the antique. Hence it was compulsory to retain and follow up the highly developed smith-craft of the middle-ages. The changes which appeared were due principally to the outward changes of form and fancy, to which art as a whole had to adapt itself. Side by side however with the new, the earlier traditions held their ground for a long time, so that in iron-work, Gothic details are by no means rare

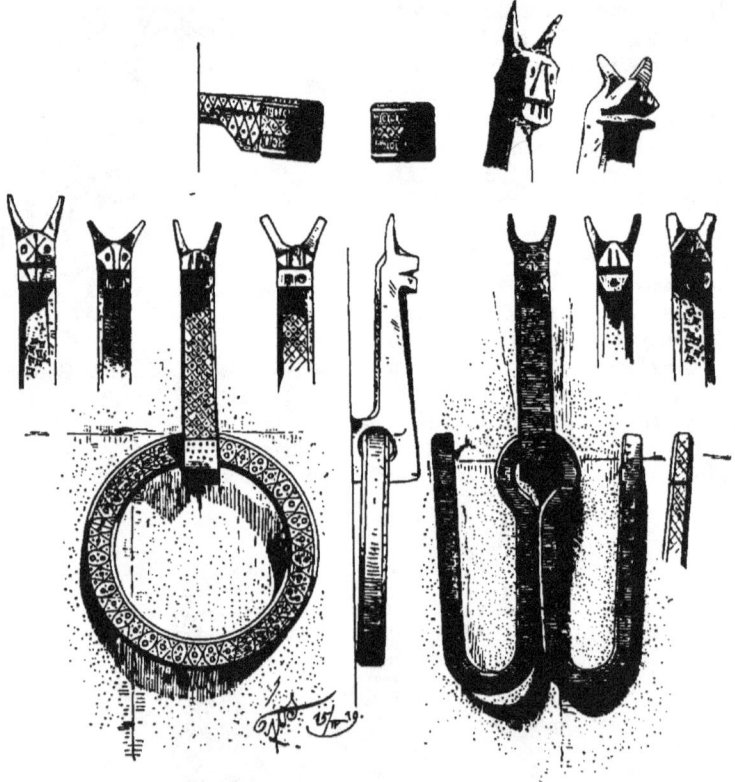

Fig. 51. Link-holders and horse-rings in the courtyard of the Bargello in Florence. 15th century.

until, and even beyond, the end of the 15th century. Fig. 49 shows a portion of a large bracket belonging to this period. The Gothic filling of the spandrel stands in contrast with the outer scroll-work which already half belongs to the renaissance style. In Fig. 50, *a*, *b* and *c* are shown three keys of which the first belongs to the 15th, the second to the 16th and the third to the 17th century. Whereas

HISTORICAL DEVELOPMENT OF THE ART OF SMITHING. 63

the first *a* is strictly Gothic; the second *b* still Gothic in its details; the ornament of the third *c* is already "baroque"; the fundamental form being in all three the same.

In Italy, where the Gothic never secured a real footing, or, as Semper puts it, where its principles were neither recognised nor understood, Gothic models were not forthcoming for the use of the smith in the same degree as in France and in Germany. The wrought-iron work of the Italian renaissance is consequently peculiar to itself, with Oriental, old Italian, Byzantine and even antique reminiscenses. The link-holders, cressets for burning pitch, horse-rings and door-knockers of Italian palaces are often remarkably simple in appearance; their ornamentation being frequently flat, and produced by geometrical punchings (see Fig. 51). Richer designs sometimes take an architectonic character which is far better suited to stone than to wrought-iron work (see Fig. 52). Late Gothic iron work both in France and Germany however was being equally forced into the same direction as is shown by Fig. 53.

With the further development of Italian renaissance a much greater freedom in form soon became apparent. Correct form is found in the organic volutes and tendril-like ornament (see Fig. 54); the addition of grotesques and emblems led to creations which were both rich in fancy and charming in effect. Speaking generally, Italian smithing retained a tasteful simplicity, without overloading, whereas in the more northern countries far greater richness was developed.

The crisp and tangled forms of the late Gothic followed on French and German soil designs which were certainly clearer and nobler in detail, but which, on the whole, produced a no less

Fig. 52.
Wrought-iron lantern, Florence. 15th century. (After Semper.)

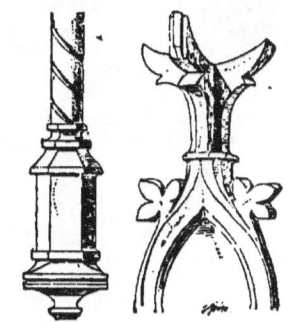

Fig. 53. Details of Gothic wrought-iron work.

Fig. 54. Altar rail at Santa Maria degli Scalzi in Venice.

Fig. 55. Circular grille in Augsburg.

rich and opulent effect (see. Fig. 55). The field open had become much wider. New objects were introduced, such as panels, door grilles, brackets with guild-shields, and tavern-signs, gargoyles, font-cover-brackets, reading-desks, wash-stands, towel-holders, weather-cocks, grave-crosses and finials, as well as utensils of the most varied kinds.

Ornamentation changed greatly in style owing to the changes in connexion with wood-work. In the place of the tongued and grooved work of the middle-ages, the joinery of the renaissance was framed and mortised. This brought about the disappearance of the long strap hinge, which was replaced by the dovetail or swallowtailed butt hinge, all parts of which could be made to serve decoratively (see Fig. 56). With regard to locks and keys a notable apparatus is introduced on the scene; curious alike in respect of external finish and as regards the mechanism. Whereas we now prefer the simplest and safest locks and the smallest keys, it would seem that then exactly the reverse was the case.

The production of weapons reached the highest perfection during the renaissance. Those of defence and offence, armour for man and charger were of the most solid and luxurious description. The incrusting and covering with gold and silver, the niello and engravings, the etching, embossing and fretting rendered them art triumphs. These processes, some of which were old and derived from the East, were further developed in the new taste applied in new ways and, finally, these purely armourers arts were adapted in greater or lesser degree to architectural smiths' work. Above all the

Fig. 56. Butt hinge. German renaissance.

glinting embossings of the armourer came especially to other uses. The principal centres of the armourer's art, such as Nuremberg, Augsburg, Innsbruck, Munich, &c., also became those of the general smithcraft. As with weapons and armour so also with grilles, embellishments and utensils the designs and drawings were made by celebrated artists.

Characteristic of renaissance grille work are the bars scrolled into volutes, the numerous instances of threading or interpenetration, the hammering of the ends into flat ornaments in the form of grotesque masks and fantastic animals, and furthermore, the free endings in the forms of conventional flowers (see Fig. 57 and 58). The flowers especially are among the finest features of the smithing of this period (see Fig. 59). The bars often received a sort of profile

Meyer, Smithing-art.   5 by being forged in swages resembling-knops and mouldings turned by the lathe (see Fig. 60). Round iron above all came into especial favour. The cutting of the acanthus leaves was excellent and simple, recalling the antique. The open elegant tendril-like ornaments were enhanced in richness and effect by embossing and lining, or etching. Colouring was resorted to either completely or else in combination

Fig. 57. Iron grille. German renaissance. 16th century.
The Ammerling collection, Vienna.

with partial fire gilding. Combinations of wrought-iron with brass and bronze came into use, especially in Italy; thus key bows are sometimes of brass or bronze while the remainder is of iron. In the middle and notably at the close of the 17th century a material change of style began to be felt, which will be treated of in the next chapter.

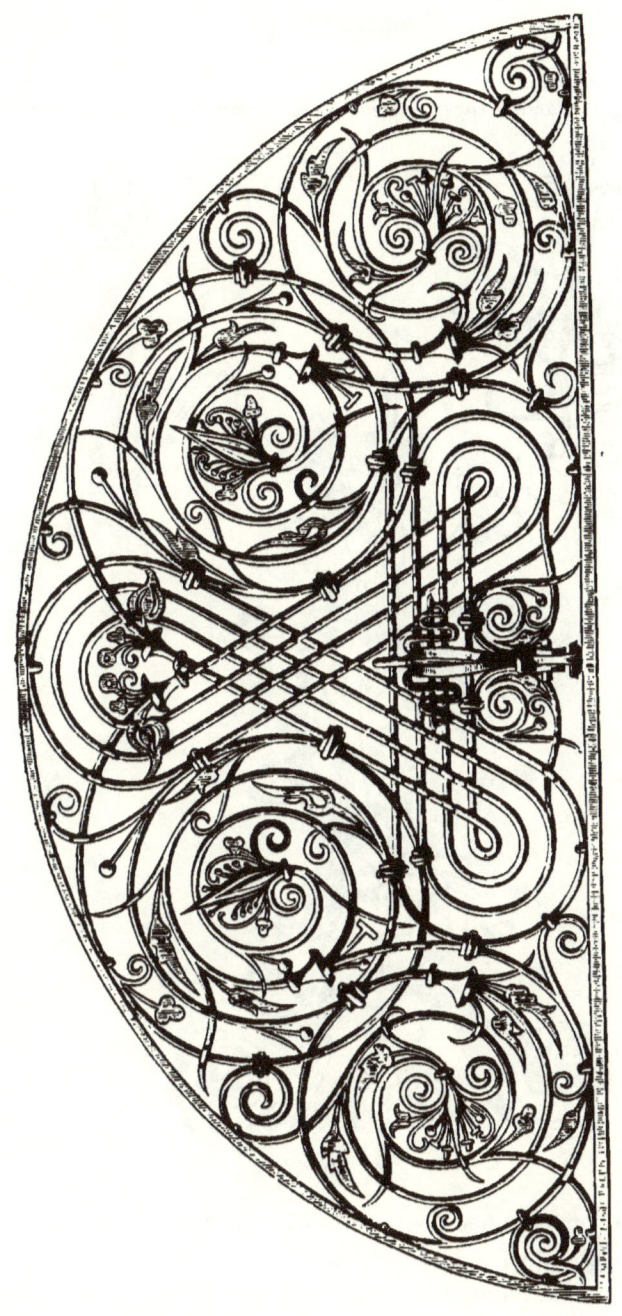

Fig. 58. Door grille in the Nuremberg Rathaus. About 1619.

68   SECTION III.

A retrospect of the renaissance period gives the following results: While the middle-ages raised art smithing, from the constructive

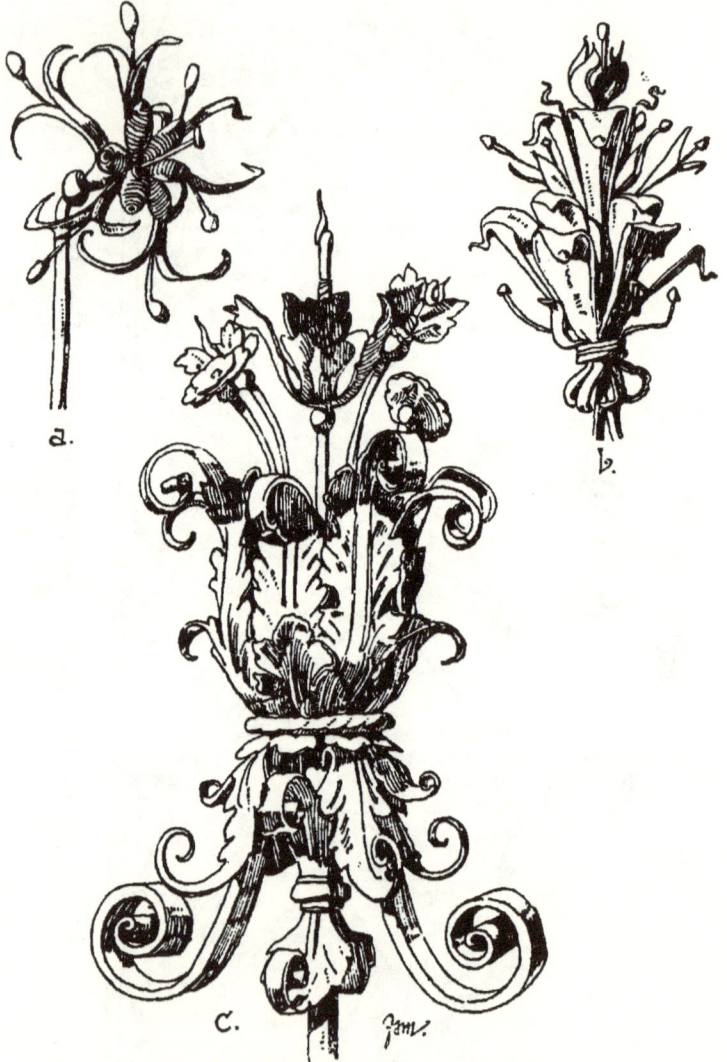

Fig. 59. Wrought-iron flowers. German and Belgian renaissance.

point of view, to the highest point, it was reserved to the renaissance to sweep away the formal degeneracy of the late Gothic, to bring about

HISTORICAL DEVELOPMENT OF THE ART OF SMITHING.

simplicity and refinement, and finally endow the art with the highest perfection of beauty of form of which it is capable. The renaissance also had the privilege of greatly popularising the armourer's craft and of opening up the widest fields to wrought-iron work generally. This period materially enlarged the field of operations, especially in ornamental work and small work, by introducing a general application of embossing, engraving, etching, inlaying, and gilding. The renaissance divided the work of the smith into distinct guilds with advantageous results.

It enjoyed material advantages over the middle-ages, inasmuch

Fig. 60. Sconce. German renaissance.

as better material, in the shape of bars, sheets and wires, was already obtainable. The introduction of iron-casting is also due to this period, though it could in its primitive stage in no wise compete with wrought-iron, and was in fact limited almost exclusively to fire backs and stove plates.

4. THE BAROQUE PERIOD.

The "baroque" and "rococo" periods have long been regarded as periods of decadence following upon the renaissance, and consequently treated with proportionate contempt. One is more tolerant of late. It is now, on closer examination, admitted that they, too, have their proper peculiarities and good points, among which art smithing takes a foremost place. One has become accustomed to regard them as independent styles. The present manual adopts this view all the more readily as its subject specially justifies their separation.

The learned are by no means in agreement as to whence the name "baroque" is derived. The term "baroque" is commonly understood to mean "oval, distorted, &c.", in so far applicable to this particular style as the adpressed, squeezed together, volutes form distinguishing features of it (see Fig. 68). This style is specially emphasised in architecture in the buildings erected for the Society of Jesus, whence it is not uncommonly called the Jesuit style.

The transition from true renaissance to baroque was, naturally, no sharper, and is far less defined, than that between mediæval and the renaissance.

The influence of the new style upon the smiths' art is principally that described below. The pompous taste of the time which verged on overloading in architecture, obtained a hold on this branch of industry. In technique it stood at its highest point, but became even more refined and eclectic in application. The first aim was to obtain great and sumptuous effects; hence it was used on a larger scale, and in relation to brass and bronze. Round iron gave place to rectangular and especially to square iron. The method of threading bars through each other, or interpenetration, gave place to halving and oversetting. Forgings applied on sheet iron backings (see Fig. 38), became more freely used. Bars were often bent into angles and formed peculiar geometric interweavings (see Fig. 61). The contour of leaves became bolder. Leaves and volutes were scrolled forward beyond the plane of the grille towards the spectator (see Fig. 62). Moulded iron was more used and came into favour for crossties and for developing forms corresponding with the open-work pediments found in architecture (see Fig. 63). Rosettes, knops and acanthus husks were used more profusely. Front and back elevations differed materially, *i e.* the application of decoration to one side only found favour. Flowers became more naturalesque in style. Wreaths and festoons came into vogue. Certain parts of the grille were treated as back-grounds, and filled with narrow crossing rods, ornamented with small rosettes at their intersections (see Fig. 64 and 68, *b*). Crowns, often far too large, cartouches, initials and coats-of-arms wrought in sheet-iron, did more harm than good. Little balls and rings were placed as connexions where scrolls and bars did not come into direct contact (see Fig. 65). With heavy iron for the constructive parts, slighter metal served for the ornament, and while renaissance grilles were often made out of one kind of bar, the baroque grille frequently required half-a-dozen

Fig. 61.
Panelled grille. Baroque.

HISTORICAL DEVELOPMENT OF THE ART OF SMITHING. 71

and more different sections of bar-iron. While the middle-ages and the renaissance aimed at producing uniform effects by even distribution, the baroque concentrated its rich effects in prominent places, leaving subordinate ones empty and plain in appearance and even reduced to straight bars (see Fig. 66). That grilles, to which the foregoing remarks chiefly apply, and balconies, balustrades, &c., should follow the

Fig. 62. Balcony. French.

curves and contours of buildings, often presenting not flat but convex surfaces, was necessary to fit them to the architecture. As framings to

Fig. 63. Details in wrought-iron. Baroque.

park and other large entrance-gates, and breaks in railings, architectural pilasters with their capitals and bases were reproduced in wrought-iron, and that mostly with taste and success (see Fig. 67).

Similar in treatment, although less important and striking, are the changes to be noted in smaller objects, such as ornaments and utensils. Retrogression rather than progress is shown in these fields. Much that was made of wrought-iron during the renaissance was now produced in other materials. The baroque style as shown by the foregoing, was intrinsically opposed in principle to small productions, and in this respect the result is but a logical sequence. Fig. 68 in

Fig. 64. Fan-light grille, Breslau University. 18th century.

conclusion, presents a series of details which are characteristic of the period now discussed.

A striving after pomp, opulence and grandeur, an eclectic refined technique, often it may be said inflated and hollow, characterise the smithing of this period. Its most striking efforts were produced in the service of courts and princes.

5. THE ROCOCO PERIOD.

The baroque style was followed in the 18th century by that known as Rococo, which found its principal fields in stucco-decoration, the "ameublement" and furniture of castles and palaces of the Regency following the death of Louis XIV of France, and during the reign of Louis XV. The word rococo is derived from "rocaille", meaning grotto and shellwork, and indicates certain salient peculiarities in the style and decoration. During the reign of Louis XIV the so-called pig-tail style came into vogue, often confounded with the rococo, though it is more correct to regard it as a style distinct in itself, as in comparison with the rococo, it shows a sobering down and return to symmetry and straight lines. Both styles are, however, highly decorative and are less apparent in external architecture than in interiors and furniture. They find expression in modelled and in plastic materials, and in these latter wrought-iron takes a high rank.

Fig. 65.
Wrought-iron detail. Baroque.

German renaissance smithing stood completely on ground of its own. The consequences of the 30-years-war were that in after times, if the independence of the craft remained, its peculiar taste and style were lost.

Art found patrons almost alone in princely courts, where it became impregnated not only with French virtues and vices but also with French taste. The result was that art on German soil fell greatly into the hands of French artists.

Cheerful and light, trifling and voluptuous, as the character of the society of the courts, so was the art of this period.

The strictly architectural lines of buildings were lost in decorative frame-work, and unconstrained flourish. Tedious symmetry was shown the door. Decorative effects were produced without rules or limits. The wrought-iron work of the rococo is dainty lace-work, a hazy web, that no longer recalls the firmness of the material employed but rather offers evidence of its great pliability. The grilles of the rococo leave, from the technical side, all that had gone before far behind them; but as to fitness of design, &c., opinions may

Fig. 66. Grilles in St. Ouen's church, Rouen. 17th century.

Fig. 67. Wrought-iron capitals. After Jean Bérain. 17th century.

Fig. 68. Various details of wrought-iron work of the Baroque.

greatly vary, this is not however the place to discuss the question. Method and skill in manipulation reached their climax during this period.

The applications of iron-work during both the baroque and rococo covered about the same ground. Grilles and sign brackets were the chief items in use. Door and cabinet enrichments became small and

Fig. 69. The insignia of a guild. Middle of 18th century. Royal Art and Industrial Museum, Berlin.

unimportant; they were reserved and made of bronze or brass by preference, at least where it was a question of rich effects. These materials were also preferred for chandeliers and other furniture: iron was no longer distinguished enough. The proletarian among metals was commonly used only where no other material would serve the purpose.

HISTORICAL DEVELOPMENT OF THE ART OF SMITHING. 77

Window-grilles became scarcer. As the times had become less dangerous they were no longer necessary. But, on the other hand, balcony railings and balustrades became all the more numerous. Churches and palaces were, as heretofore, provided with large pom-

Fig. 70. Panel for staircase. 18th century.

pous-looking iron gates. But, above all it was the parks which opened up a wide field for grilles and railings, as may be observed

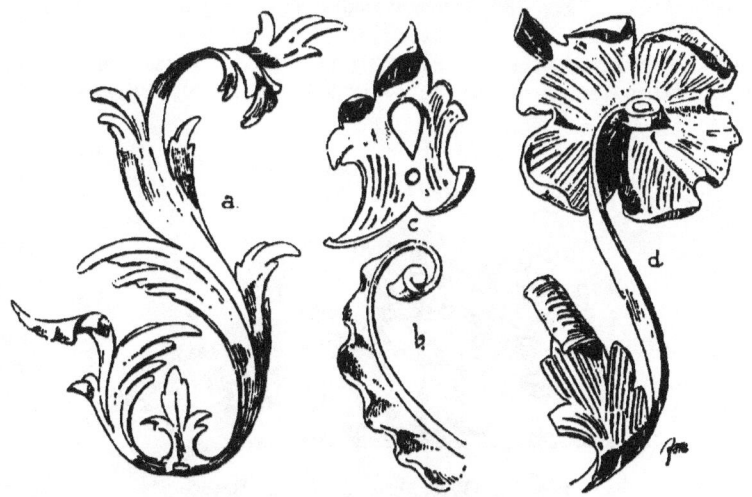

Fig. 71. Wrought-iron details. Rococo.

at Versailles, Würzburg and Schwetzingen. As regards tavern and craftsmens' sign brackets and signs for guilds, there was an increased rather than decreased demand. Wrought-iron became more popular than ever in this particular branch. Nearly every little town, every village, can still show suchlike art blossomings. The same is the case with regard to fan-light grilles and, at least in certain districts, to crosses for steeple and churchyard.

On investigating the characteristic features of the wrought-iron work of the rococo period the first thing that strikes the eye is the abandonment of symmetry already referred to (see Fig. 69). Another

Fig. 72. Wrought-iron detail. Rococo.

point is the marked avoidance of straight lines. These are only retained when the nature of the construction positively required them,

Fig. 73. Wrought-iron detail. Rococo.

or the use intended, precluded other treatments. A geometric design is only found as a rule when the work is but a poor skeleton arrangement, or where in isolated parts it serves as a reposeful contrast. In its place we have arbitrary, disordered and wild scroll-work (see Fig. 70).

Flat bars with rectangular section came into favour. Volutes

HISTORICAL DEVELOPMENT OF THE ART OF SMITHING. 79

and foliage were treated more luxuriantly and thrown into more daring relief. Acanthus foliage was, as in the Gothic, once more drawn out, and deeply cleft with peculiar outlines (see Fig. 71, *a*). Unmeaning, indefinable elements appear in the decoration (see Fig. 71, *c*). The crinkling of the foliage is also peculiar to the period (see Fig. 71, *d*).

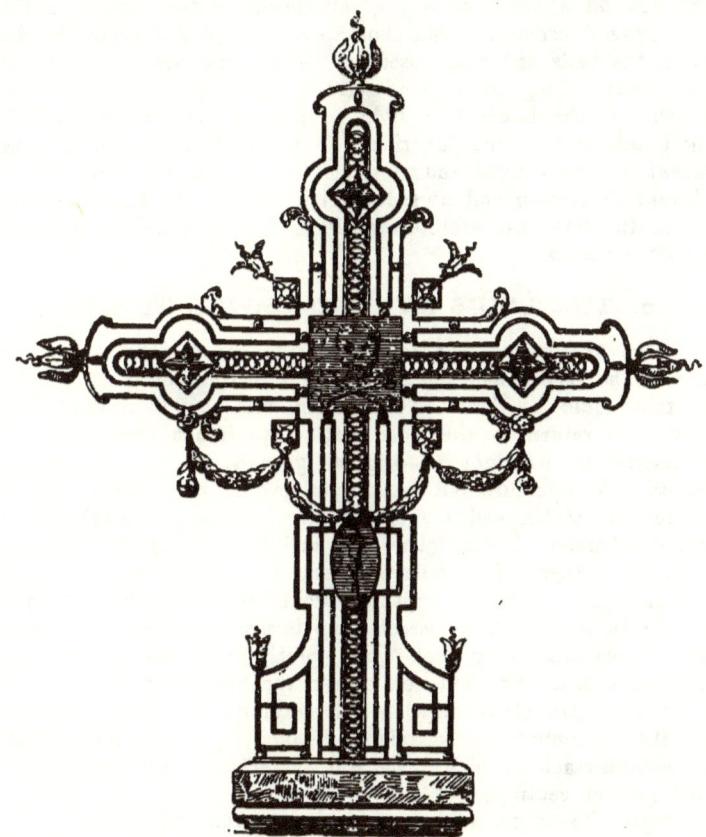

Fig. 74. Grave-cross.

It is evident that this was produced in the desire to avoid flat surfaces as much as possible and to throw more life into the work by simple means. This process recalls the "rustic" and the "Vermicelli" carving of stone in architecture and wall-decorations of the same period.

Characteristic, again, is the pleasing interspersion of naturalistic flowers and fruits. Sprays, garlands and festoons fill up every empty space and illustrate the spirit of the craftsmen of the period in an eminent degree (see Fig. 72).

What would the art-smith of the 12th century have said could he have seen such a specimen of work as this! Then again, and these are very characteristic, meandering, interwoven, wavy, and similar borderings on a small scale are introduced in the winding outlines of the general ornament, and thus apparently played about the same part as the balls and rings used in the boroque (see Fig. 73). But these things belong to a comparatively speaking, later date. They are either of the Louis XVI style, or link it with the rococo. The highest development and luxuriance in point of technique, the abandonment of architectural and constructive rules, in favour of arbitrary, exuberant decoration and a gradual giving up of the smaller kinds of work in the service of architecture, form the characteristic features of the rococo period.

6. THE LOUIS XVI AND EMPIRE STYLES.

These two periods are condensed into one for the sake of simplicity. The rococo had reached the culminating-point: all had been done that could be done. The reaction was now in favour of simplicity, af a return to the older ways. Not much good came of it. The movement was not of original growth; it was the result of necessity. Architecture fell back upon would be classical, but ofttimes tedious, styles, and it was, of course, the same with art-smithing. Antique vitruvian scrolls, interwoven and flowery borders were introduced into grilles. The foliage became stiff and small; attenuated laurel garlands and wreaths with manifold bows and ribbons, enframed plain elliptic shields. Grave-crosses and tavern signs became extremely sober. From afar they look quite pretty and lead one to expect something satisfactory; but, on nearer approach, they are commonly not worth the trouble of drawing. Everything continued to degenerate from the beginning of the reign of Louis XVI until the Philistine stand-point is reached which, in general, characterised the second quarter of the present century.

Figure 74 shows a grave-cross dating from this period.

7. THE PRESENT DAY.

The breaking out of the French revolution seems to have been a turning-point in the history of art just as it was in the history of nations. The waves of this mighty movement carried away the prevalent styles as it reduced many other things to mere ruins. It is true that the revolution did not directly affect the majority of races, but it materially broke down French influence, or, where that remained, new roads, nevertheless, became opened up. The time of the Empire and the rest of the, artistically speaking, almost dormant first half of the 19th century, failed to give any fresh impulse to the nearly defunct craft of the smith. The Empire was too classic and sedate and the remaining time was too prosaic for this. The technical progress made during the said half-century was devoted to ironworking and ironsmelting and utilising cast-iron to the utmost, and to attempts, by means of improvements in iron-casting, to render it the medium for art work in place of wrought-iron. And for a long time it seemed as if such would be the case. However, with the re-awakening and revival of art handicrafts during the last decade, which have been brought about by schools for art industries and by museums, the conclusion arrived at is that, in spite of all its advantages and notwithstanding the high perfection to which it has been brought, iron-casting must always remain suited to particular purposes only. The fact that castings aim at the exclusion of all undercutting, that the skin of cast-iron is of such a nature as to render the subsequent manipulation a matter of great difficulty, has brought about the conviction that wrought-iron offers, in almost every respect, a material which is far more plastic and, decoratively speaking, effective than cast-iron. And thus it has happened that the latter, in spite of the advantage of cheapness, has already been driven out of many a position where it had, to all appearances, taken firm root. The originality of handwork has triumphed over mechanical factory-work. Even when the smith's hand-wrought object of art is repeated dozens of times, each individual copy retains some individuality, and this is exactly what cannot be the case with castings. The contrast is like that between the music of an orchestra and that of an orchestrion, if such a far-fetched comparison be permissible. A certain domain will, and rightly, always be left open to iron-casting — such for instance, as for stoves, the pillars and bearers used in building — and in commercial art fields it will remain a cheap substitute for the results of manual labour.

When once the dead point was overcome the smith's art revived with great rapidity, far more quickly indeed than would have been the case if the revival had taken place two or three decades later, for the reason that the old traditions were not altogether forgotten. Here and there was still to be found an old master-smith, who had in his apprentice-years obtained a thorough and practical training, so that, comparatively speaking, in a very short space of time the long dormant exercise of the art awoke and the desired skill was brought once more into activity. And, at present, in the days when this manual is being written, modern smithcraft is producing everything possible to it. What was formerly made can be made now, even if it has not yet been made, because the general appreciation and enlightenment of the age is not sufficiently advanced to inspire the public to give such orders, though that stage may be attained in the course of a few years.

If we ask whether modern smithing has already found a style of its own, the question may undoubtedly be answered in the affir-

Fig. 75. Ornament by F. Brechenmacher of Frankfort on the Maine.

mative, even although superficial appearances would seem to assert the contrary. We are too closely surrounded by our present art-productions to be able to take so comprehensive and undisturbed a view as those presented by more distant ages. Let any one examine a good piece of wrought-iron work of the present day and one of former times. Will he ever mistake one for the other if he has even a limited knowledge of the matter? Certainly not, but why? Firstly, because modern industry works with very different means;

Fig. 76. Portrait in Relief of the Grand Duke Frederick of Baden, embossed in iron by Professor Rudolf Mayer of Karlsruhe.

machinery has altered and multiplied the tools; the present ability to obtain rolled-iron in so many forms, the machine-made rivets, knops, rosettes, &c. gives to contemporary smithing a modern imprint and leads to different combinations and constructions. Secondly, its field of employment has materially altered and is in part entirely new. Let us, for instance, consider lighting-apparatus. In the place of oil-lamps and candles, or, at least concurrently with them, are found gas- and electric-lighting. Now, these newly introduced sources of light require supports of materially divergent character. Thirdly, our views as to style, and external form, are different. Modern times have often been branded as without style simply because work is done in all styles, and all possible periods are drawn upon for models. This revival of the various styles of former days, this universal many-sidedness, this adapting of a medley of styles to modern requirements, will alone suffice to set a peculiar stamp on the style of to-day.

Two distinct ways are clearly recognisable as those by which the smiths' art has reached its present state of efficiency. One of these is found in the circumstance that from the plain railing and grille work, which, until a few decades back, answered all purely practical purposes, the rich and elegant door grilles and fan-lights, &c. now to be found were developed by a very gradual addition of ornament. The other way was by directly imitating and copying old models. It has become a fashion to copy as exactly as possible, retaining both their good and bad points, the known and recognised objects which are stored up in our museums, and to sell the same to lovers of art and connoisseurs. This course, as compared with the first, has many dangers, nevertheless it must be recognised as serving as a means to an end. If the administrations of our museums, schools, and associations of art and industry unite to raise the art of smithing to she utmost point of their power, they will be but fulfilling their duty and obligations. But the basis of all success lies in the workshop and it is deserving of the highest recognition that masters, such as Puls, as Kramme in Berlin, as Milde, as Gillar in Vienna and others made it their task already in early days to restore smithing once more to its right position.

In 1887 the Baden Art-Industry Association offered a prize for finished smiths' work and collected the competing exhibits, together with various things connected with the art in a special exhibition. About 60 exhibitors, hailing from all parts of Germany, sent in over 300 objects made by them, and some of the work was of the highest excellence. This exhibition was exceedingly interesting and exactly calculated to afford a picture of what the modern smiths of Germany could produce. The picture was most pleasing and proved clearly that the efforts made on all sides to return to genuine smithing and to restore the craft to its original and early state, were being crowned with success. A new feature, which is likely to have a future, was presented in the shape of articles made of embossed and forged delta-metal, a kind of bronze of most beautiful colour.

HISTORICAL DEVELOPMENT OF THE ART OF SMITHING. 85

This is not the place to discuss the exhibition in detail. The most important articles were photographed and reproduced in phototype for publication.*)

We illustrate this section of the historical development of the

Fig. 77. Door sign by the Smith Bühler of Offenburg.

art of smithing with two of the objects from this exhibition

*) Modern German Art-Smithing in 7 parts, each with 6 plates in phototype at 5 marks each. Bielefeld's library (Liebermann & Co.), Karlsruhe.

Fig. 78. Smiths' work by Cassar of Frankfort on the Maine.

represented in autotype. One is an ornamental piece of detail, a tendril by F. Brechenmacher of Frankfort (see Fig. 75). This prize work shows extraordinary boldness in smithing and can, incontrovertibly, be ranked as high as any work of the last century. The second object is (see Fig. 76) a profile-portrait in relief of the Grand Duke Frederick of Baden. With this piece of work, which was not for competition, Professor Rudolf Mayer of Karlsruhe, who chased it, showed what a high degree of artistic capability there is in wrought-iron, and how much can be done with it by duly skilled hands.

Since the first appearance of this book the smithing-art has not been idle. Brechenmacher's laurels gave such an impetus that good smithing is now nearly everywhere actively cultivated. The number of real art-smiths has so increased in this short time that they cannot all be named here. The space available for illustrations is, moreover, too limited to permit of showing specimens of all the various kinds of objects for which wrought-iron is suitable. Figures 77 and 78 furnish examples which accident has placed at the author's disposal.

SECTION IV.

THE PRINCIPAL FIELDS OF ART SMITHING.

1. GRILLE WORK AND BALUSTRADES.

As shown in the preceding section, the fields for the application of wrought-iron appear under different periods and styles to have been subject to certain changes and vicissitudes. Among the objects which were almost without exception and in all times made of wrought-iron, balustrades and closing grilles may be counted. This at least holds good from the time of the middle-ages up to the present day. The antique, however, seems to have made no use of wrought-iron railings, inasmuch as no reference is made to them by old writers and no specimens have come down to us. Where railings are depicted on vases or on sculpture they are of such a nature as to exclude the idea of wrought-iron having been the material employed.

From the early middle-ages we find, on the contrary, that wrought-iron grilles or railings were introduced, at first, as was natural, in simple and inartistic form, windows and other orifices for light were small and often consisted of loop-holes only, so that grilles could under no circumstances assume large dimensions. The finishing-off of parapets, balconies and such-like was executed in massive stone-work, so that in these again wrought-iron could play no part. Then again, for doors and gate-ways grille work hardly even entered into consideration, as the times demanded them to be closed with heavy wooden folds studded with iron for defensive purposes. The first attempts at artistic grille work are probably to be found in fire-screens and articles of furniture of a similar nature. Already in Fig. 42 such a screen is shown. It is wrought out of the piece and dates, to all appearances, from the 13th century.

THE PRINCIPAL FIELDS OF ART SMITHING.

The grille work of the earlier middle-ages, setting aside work destined for interiors, was intended for protection rather than ornament. Such work is strong and massive and menaces the unauthorised trespasser, by the pointed bristling ends, with impalement or with injury to body and raiment as the penalty of attempted intrusion. This barbaric direction is opposed to aesthetic feeling and reminds one to a certain extent of our modern fencings of spiked wire. A few examples of such work, after Viollet-le-Duc, are shown in Fig. 79.

Later on, in the Gothic period, grilles became more frequent. They were used in

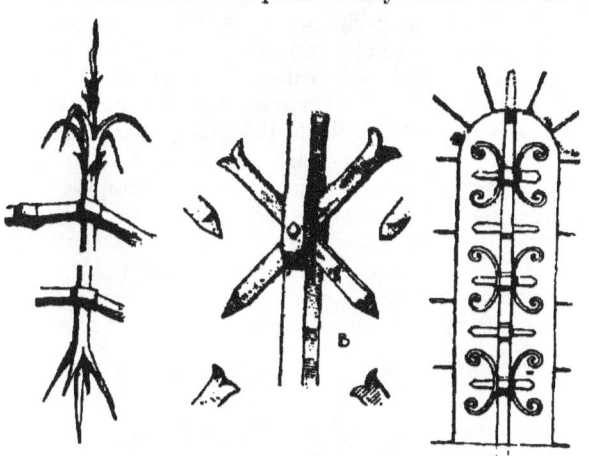

Fig 79. Details of grilles of the middle-ages, from Viollet-le-Duc.

the churches to close in chapels, altars, monuments, &c. The wells in cloisters and courtyards of castles were often railed-in. Grilles gain at the same time in richness and elegance. Altar and chapel grilles become comparatively high, much above the level of the eye. The prevalent form consisted of a row of perpendicular bars fastened into a few cross bars of iron which formed together the frame-work, and between which the pieces to serve as decoration were placed. (Compare Fig. 43 on p. 57 with Fig. 80 on this page.) The upper ends of the bars were generally finished off in the form of fleurs-

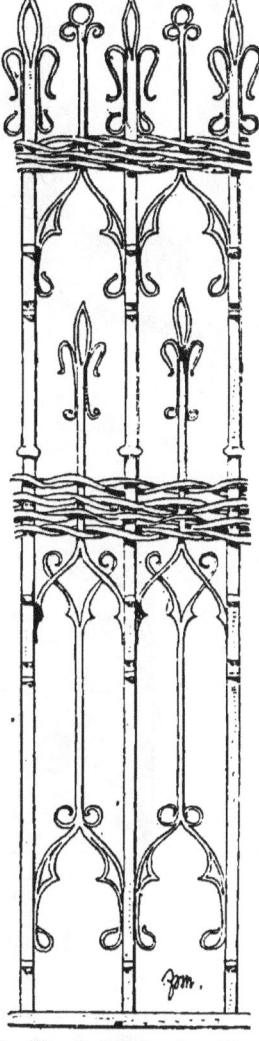

Fig. 80. Gothic tomb railing in the church of Breda. 15th cent.

de-lis (see Fig. 80 and 81). Square-iron was almost invariably used. The ornamental parts were mostly made of thinner flat-iron. The vertical bars placed anglewise, passed through holes either punched or chiselled in the cross rails.

In the late-Gothic period another style of grille is found, in addition to the above. As in carving, geometric patterns were produced in flat iron which recall the wall-painting and textile art of the same period, and for the most part worked out evenly without any special right way up. This form of grille work is very effective, requires only simple work and is especially suited for large pieces.

Fig. 81. Cresting of grille in the Cathedral, Toulouse. 15th cent. From Viollet-le-Duc.

The task of finishing the top of a grille of this kind is far more difficult than in the first-named kind, in which every vertical bar had its free end. As a rule, only the main bars of such even surfaced geometric grilles, which divide them into regions, finish in lilies or other flowers, &c. An independent cresting having no connection with the other parts was sometimes placed over

Fig. 82. Geometrical designs of grilles.
a. Chapel screen in the Cathedral at Perugia.
b. From the Campo Santo at Santa Croce, Florence.

the geometric areas. Fig. 82 shows two geometric railings of this kind, the basis being in both the quatre foil.

The employment of grille work increased further during the renaissance, and this not only in churches but also in private homes and in public buildings, such as Townhalls, Exchanges, &c. Low balustrades are also to be found along with the high grilles. Staircases and flights of steps as also the approaches to chancels and platforms, offered welcome opportunities for the use of balustrades. Oriels and alcoves were often divided off from the principal appartment by grilles. The window-openings and fan-lights were richly grilled.

Square and angled iron became replaced by preference with round iron. In order to avoid needless repetition, the reader is referred back to what is said in section 3 with regard to the remaining changes in styles and periods of work.

Renaissance grille work may be classed under three groups. First the early style of grille formed of bars was retained, together with the same kind but with added ornament, and such modifications as the new style demanded. Fig. 83 gives two examples of this class. The one on the left still exhibits Gothic reminiscences, although it belongs, both in respect of time and in its foliage, to the renaissance.

The second group consists of the further development of geometrical tracery, carried out in the flat. In combination with the predominant quatrefoil designs of the Gothic period are to be found numerous other varieties, such as the diagonal trellis in which parallel bars cross each other obliquely. Varieties belonging to the second group are shown in Fig. 84 and 85, the last being taken from the author's "Manual of Ornamentation".

Fig. 83. Renaissance grilles.
a. Grille closing a chapel in Freiburg cathedral, Bad. End of 16th cent.
b. From St. Mary's church, Dantzig. Beginning of 17th cent.

The third group comprises the Panels consisting of framework filled in regularly with some specific pattern. Inasmuch as these rarely occur in the styles of the middle-ages they may be counted as belonging to the innovations introduced under the renaissance. The form of the filling is of course governed by the position it is to fulfil. Along with the vertical and horizontal rectangular filling are to be found the square, the circle, the ellipse, the stilted and depressed arch, the semicircle and the lozenge which for the most part form the basis of the pattern. The right angle, square, circle and ellipse are principally used in door and window fillings, while the stilted and depressed arches, and the semicircle, are adapted to fanlights. Lozenge-shaped panellings and those with irregular angles are found almost only in staircase-balustrades, where the slanting position following the steps necessitates such change of form. Besides these are occasionally to be found all sorts of arbitrary forms, of which the regular and irregular polygons and the spandrel forms are entitled to special mention.

Two principal features serve to form a basis for the classification of panellings. The design may show a distinct top and a bottom end, when it is simply symmetrical. Or again, the ornament may be developed from the centre in all directions in equal proportions, when it is bi-symmetrical or repeating. In the first case we have an upright, and in the last a central filling. The middle of a central filling is not uncommonly distinguished by a rosette. Little divergencies from absolute symmetry and perfect regularity often occur so far as to affect detail only without disturbing the regular effect of the whole.

Fig. 84. Repeating designs for wrought-iron panels.
a. At Santa Maria sopra Minerva, Rome.
b. From Venice.
c. At the Ospedale Maggiore, Milan.

In filling squares the natural lines which govern the ornament are the diagonal, vertical and horizontal lines. The square thus divides into 8 equal triangular spaces filled with the ornament. Fig. 86 shows some square panels with ornament belonging to the renaissance period.

Inasmuch as the circle presents no ready lines of subdivision, the filling-in is usually by radial lines dividing it into any number

Fig. 85. Various Repeating designs in wrought-iron.

of equal parts. The commonest subdivision is into three, four, six and eight (see Fig. 87, b). Sometimes the circular panel is filled wholly with an upright, symmetrical design (87, a).

Elliptical window-openings, sometimes erroneously called oval, appear sometimes with upright and sometimes with horizontal axes. The large and small axes suggest the natural lines whereby the ellipses can be divided into four equal parts (see Fig. 88).

It is the same with panels which have somewhat the same form

Fig. 86. Square panels in wrought-iron.
a. and b. French Renaissance. c. and d. German Renaissance.

as an ellipse with elliptical ends and with or without parallel sides (see Fig. 89).

The stilted and the depressed arches and the semicircle, which are used in fan-lights, are sometimes filled in with an upright, symmetrical ornament, or, and this is specially the case with semicircles, with many radial subdivisions, but in such cases, in order to avoid the inelegant meeting of the radial bars at one centre, a smaller open semicircle, or one which is ornamented independently, is introduced. It also happens occasionally that the semicircle is divided into separate zones, each one of which is treated as a band of ornament

THE PRINCIPAL FIELDS OF ART SMITHING. 95

in itself. Fig. 90 shows a depressed arch; Fig. 92 represents two stilted arches and Figures 91, 93 and 94 are specimens of semicircular fillings.

Diamond or Lozenge shapes are less often used as a panel because this shape of window-opening so rarely occurs in architecture. They are more often to be found in joiners-work and doors. The ornamentation is either diagonal from a central point, or it is symmetric and vertical. It is not uncommon in grille work formed by the crossing of diagonal bars, to find the logenze-shaped subdivisions filled in, in order to produce more variety in what would otherwise present a monotonous appearance. When the distribution is judicious a good effect is produced. Fig. 95 shows some specimens of lozenge-shapes filled in.

Similar in treatment to the lozenges, which are nearly always placed with the axes vertical and seldom horizontal are the squares placed on the angle. The ornamentation is almost invariably from the centre (see Fig. 96).

In Regular polygons the angles and the centre points suggest points of departure for lines passing through the centre and dividing them into equal spaces, a radial arrangement again being most effective (see Fig. 97).

Fig. 87. Circular panel, German Renaissance.
a. Saint Saviour's, Prague.
b. From an Augsburg house. 1550.

The Rectangular is by far the most frequently used form of frame-work, as will be readily understood. This is used, both vertically and horizontally, according to whether the wings of the design

SECTION IV.

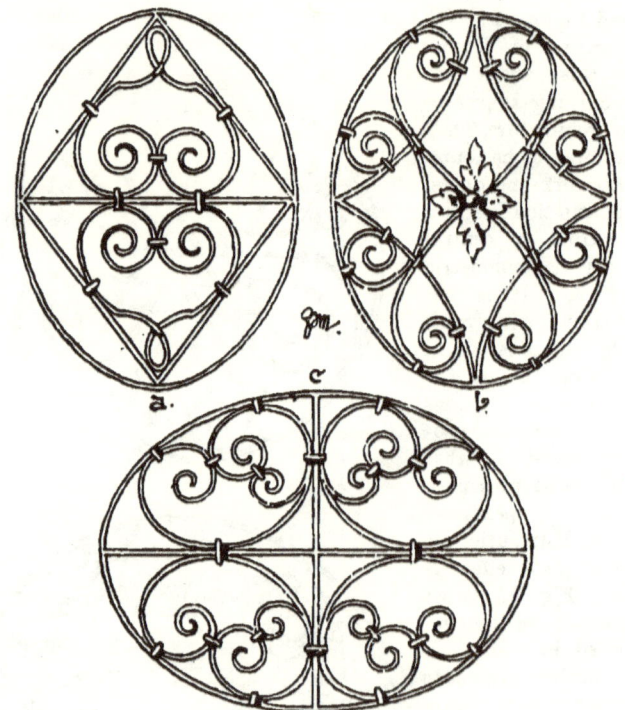

Fig. 88. Elliptical Panels.
a. From Pisa, Via S. Martino. b. From Verona. c. From Venice.

Fig. 89. Wrought-iron Panels from Venice. Italian Renaissance.

are higher or broader. Rectangles are as well suited to a central as to an upright symmetrical design. In centred designs the transverse lines uniting the outer bars of the frame at their centres form the principal guiding lines. Less useful are the diagonal lines as the angles are unequal. The rect-angle is almost universal for door, window, and parapet panellings. Fig. 98 shows horizontal rectangular panels, one with an upright, the other with a central filling, while Fig. 99 shows perpendicular panels, one with an upright, the other two with central fillings.

The Oblique panel or Rhomboid is only used for staircase-balustrades, the rise of the steps necessitating this form. The fillings are generally arbitrary and irregular (see Fig. 100, *a* and *b*). If such a panel approaches the lozenge shape, diagonal lines may be found useful (see Fig. 100, *c*). The horizontal cross-division in Fig. 100, *d* is very remarkable.

The staircase-balustrades already referred to consist mostly of many panels placed alongside of each other. The same arrangement is sometimes seen where large closing grilles are formed of a number of rectangular panels placed together. In such case the principle of an unvaried series of repetitions is abandoned in favour of another whose mass is formed of an assemblage of separate panels. As the frequent repetition of one and the same filling pro-

Fig. 90. Fan-light. German Renaissance. Villa Bergau, Nuremberg. 16th century.

Fig. 91. Fan-light. German Renaissance.

Fig. 92. Fan-lights. a. Venice. b. Innsbruck.

Fig. 93. Fan-light. Italian Renaissance.

THE PRINCIPAL FIELDS OF ART SMITHING. 99

Fig. 94. Fan-light. Italian Renaissance.
a. At S. Giovanni in Monte, Bologna. b. At Sta. Maria Formosa, Venice.
c. At via Garibaldi, Perugia. d. At. S. Antonio, Pisa.

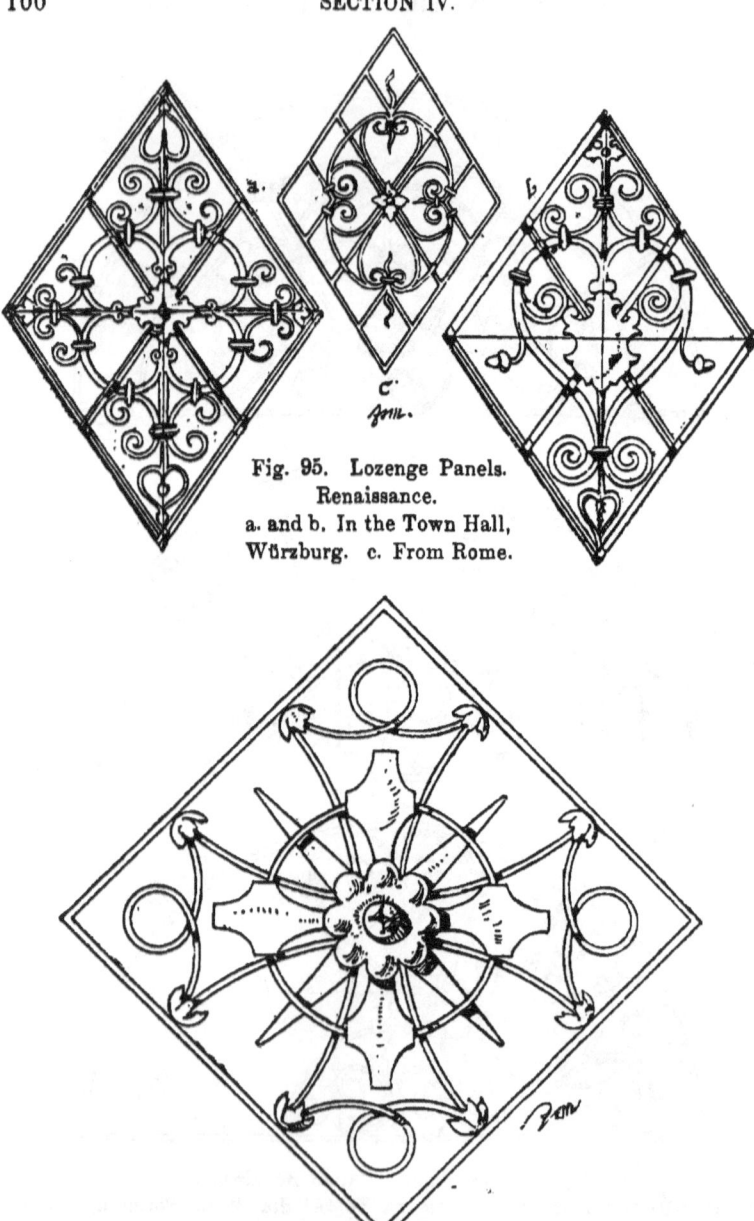

Fig. 95. Lozenge Panels.
Renaissance.
a. and b. In the Town Hall,
Würzburg. c. From Rome.

Fig. 96. Quadrate Panel, from the Campo Santo, Bologna.

duces a somewhat monotonous effect, variations of a given design are introduced or quite as often designs entirely differing from each other.

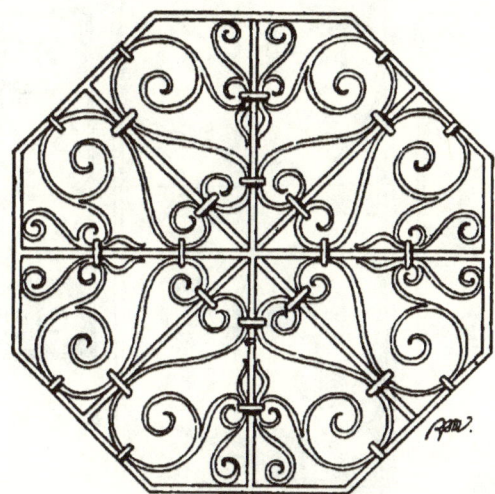

Fig. 97. Octagonal Panel in S. Petronio, Bologna.

Fig. 98. Rectangular horizontal Panels. Renaissance.
a. From Schlettstadt. b. From Italy.

This free treatment is shown in Fig. 125, which will be discussed more closely under the heading of doors and gateways. But Fig. 101

Fig. 99. Upright rectangular Renaissance Panels. a. From the Abbey of Strahow, Prague. b. In the Church of St. Blasius, Mühlhausen in Thuringia, mid 17th century. c. From Padua.

Fig. 100. Staircase Panels. a. and b. From the house "Zum alten Limburg", Frankfort on the Maine. 16th century. c. and d. From the Cathedral of Thann in Alsace. 16th century.

Fig. 101. Grille surrounding the Cenotaph of the Emperor Maximilian, Innsbruck.

already shows a very finely designed and executed grille which presents the same kind of variation. It represents the celebrated railings round the monument to the Emperor Maximilian in the Franciscan Church at Innsbruck and shows among other things what

Fig. 102. Cresting to the Augustus Fountain, Augsburg.

effects may be produced by rhythm and contrast, inasmuch as the spaces with geometrical ornament alternate with those in which organic plant forms predominate. The same example also shows how such grilles may finish in crestings or terminals. The reader will not have failed to notice that various details forming parts of this glorious specimen of wrought-iron work have been used in preceding illustrations.

Fig. 103. Side-sections of window-railings at Verona.

Fig. 102 shows part of the cresting to the Augustus Fountain at Augsburg. The grille work and balustrades of the baroque have been discussed in the third section in connexion with changes in style. As the smiths' art of this period was exercised principally in the service of princely courts, grilles and gates for gardens and parks take the leading position. Church exteriors and interiors also present specimens of magnificent grilles. In the architecture of palaces and the residences of the wealthy, balconies are next in importance, and the window grilles, which now take peculiar forms. These window grilles are often bowed

THE PRINCIPAL FIELDS OF ART SMITHING. 105

out in their lower part so that the outlookers might have a wider range of view. This bulging-out converts the window grille into a sort of case or cage, the sides of which present opportunities for elegant ornamentation (see Fig. 103). These sides have for us a present interest, inasmuch as viewed horizontally they furnish ideas for consoles and wall-brackets.

Detached closing grilles often present the character of vertical bar-railings into which ornament is only introduced here and there (see Fig. 66), and this was necessitated by the great extent of the walks and pleasure-grounds they enclosed. Such railings, unless the architecture itself imposes division into separate parts, are divided at certain distinct intervals by stouter iron-uprights or by the interpolation of open-work pilasters. The tops of the individual bars

Fig. 104. Cresting of grilles at Halle on the Saale. About 1740.

are finished in the form of spear-heads, &c., and often two or more bars are combined in order to form a terminal (see Fig. 104).

Grilles with endless geometric figures, such as those common in the Gothic and Renaissance periods, went almost out of fashion, and that which has been previously said with regard to symmetrical grilles remains applicable. For instance, Fig. 105 and 106 present examples belonging to the Baroque period.

In fan-lights, too, stately work was done. The contours were partly taken from those known in the renaissance, as, for instance, the semicircle (see Fig. 107). Moreover, the frame-work was often capricious and sportive in outline, and sometimes the grille was even without any actual, definite frame, as seen in the illustration of a fan-light in Fig. 109.

The renaissance mode of composing large and complex grilles of separate panels was pursued further but with this difference,

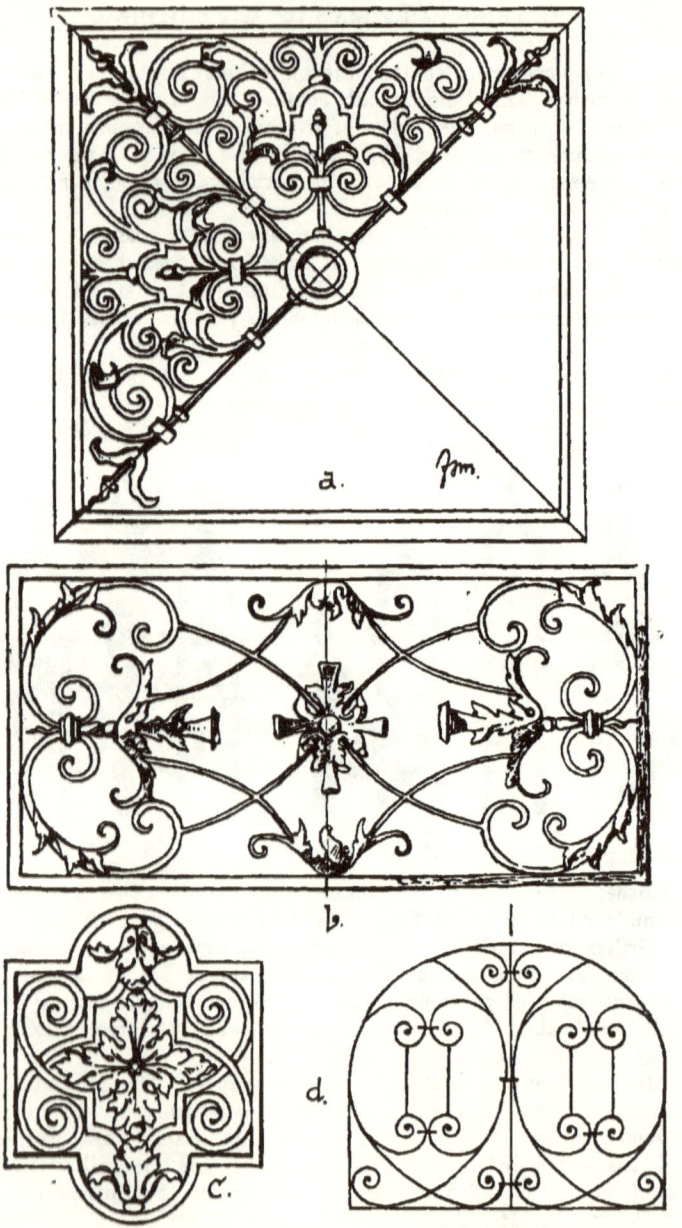

Fig. 105. Baroque panels. a. From a gate finished in Oxford in 1713. b. From a house in Freiburg, Switzerland. c. From Vienna. d. From Padua.

namely: that, as a rule, they were not built up of several equally large panels forming a whole; but of small panels alternating with large, narrow with broad, plain with rich ones, &c. The crestings belonging to them are not often materially distinct from those previously in use (see Fig. 108 and 110).

A fashion of grille, a part of whose design was made up of rods intersecting each other at right angles, so as to leave empty rectangular spaces of different sizes, whilst the remainder was composed of scrolls, rosettes, &c., must have appeared a new departure. Fig. 98 illustrates this style.

Finally, the designs shown in Fig. 112 represent several baroque panels in the then prevalent taste.

Straight lines and pronounced structural features disappeared during the transition from the baroque to the rococo period and made way

Fig. 106. Baroque Panel.

for frames of scrolls and flourishes. This indicates at the same time that panel work formed the principal feature in railings, grilles and balustrades. Bars were necessarily retained for park railings and other large enclosures, but these were combined with the arrangements and distribution peculiar to the baroque, to suit which the details of the pilaster, cresting, &c. were equally modified.

Symmetrical designs were gradually discarded in favour of asymmetrical, this was the case even with fan-lights and window openings where such an arrangement was hardly tolerable.

In Figures 113, 114 and 115 three panels are shown which belong partly to the transition from the baroque to the rococo, and partly to the latter. Fig. 70 also represents an example belonging to this period.

As far as regards the grille work of the present age, it consists partly of more or less direct copies of specimens of earlier styles, and partly — this applies principally to work of a plainer kind — of designs of specifically modern character. In the latter the work clearly shows the endeavour to produce a good and rich effect with the least possible expenditure of time and money. The set-square and compass play the chief part in the drafting of designs, and strip- and sheet-iron are frequently the only materials used in the execution. The principal fields for the employment of modern gates and grilles are garden and tomb rails, rectangular door-panels and fan-lights, balconies, window-grilles and stair-rails. The wrought-iron baluster-rails, which follow each other in regular order, and support and join the steps with the hand-rail, recall the time when iron-casting was in the ascendency. Timid attempts are also occasionally made to introduce grille work into furniture. Such experiments go hand-in-hand with the prevalent fashion, in rooms, alcoves, antechambers, bay-windows, &c., which aim at reproducing the so-called Old-german style. The space to which this manual must be confined does not admit of illustrating all in this connexion.

Fig. 107. Fan-light, beginning of 18th century.

Fig. 116 to 121, however, give some idea of the direction of the modern art.

Fig. 108. Cresting to the Hercules Fountain at Augsburg.

Fig. 109. Fan-light grille in St John's Church-yard, Leipzig. 1734.

Fig. 110. Cresting to the Hercules Fountain at Augsburg.

Fig. 111. Window grille, St George's church, Halle on the Saale. 1744.

THE PRINCIPAL FIELDS OF ART SMITHING.

2. DOORS AND GATES.

Setting aside the question of wooden doors embellished with iron-work, which can be better dealt with in the next chapter, and

Fig. 112. Patterns of Baroque grilles. a. San Martino, Pisa. b. and c. Venice.

treating only of open-work iron doors and gateways; the middle-ages present little for consideration, inasmuch as wooden doors were

Fig. 113. Fan-light grille to a house in Como.

Fig. 114. Fan-light from Innsbruck.

Fig. 115. Rococo panel from Schoenenberg, near Zurich.

THE PRINCIPAL FIELDS OF ART SMITHING. 113

Fig. 116. Modern railing, designed by the Author.

principally used at that period. Where iron gates are found in chapel, grave, and similar enclosures, the style is generally simple in character. A part of the fixed grille of vertical bars is bound together and

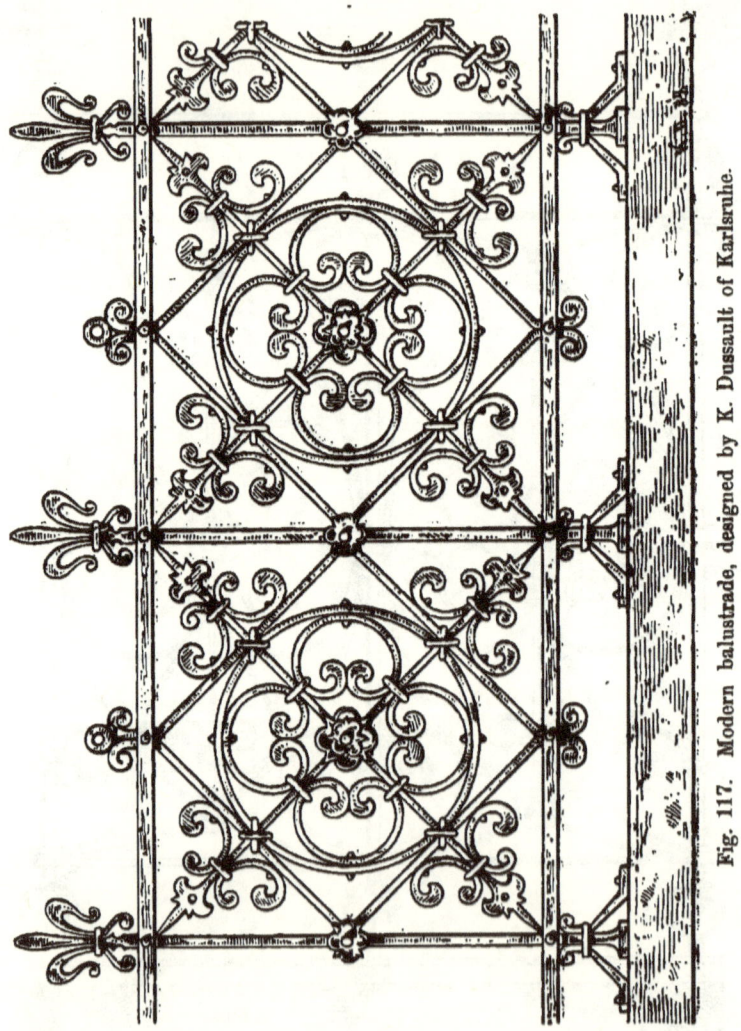

Fig. 117. Modern balustrade, designed by K. Dussault of Karlsruhe.

revolves with the aid of pins and sockets. Where grilles are formed of multitudinous repeating geometric ornaments a sufficient portion is securely fastened together and utilised as the gate. While vertical

Fig. 118. Modern tomb-railing, by Professor Th. Krauth, Karlsruhe.

Fig. 119. Modern balustrade, designed by the Author.

Fig. 120. Modern grille, designed by the Author.

bars are carried through and thus form their own crestings it is not necessary in the other class of grilles to make the gate of the entire height. When carried up to the full height, however, the gate was surmounted with a cresting of special ornamental character, or else the cresting is fixed to a horizontal bar forming a lintel so that it does not move with the gate.

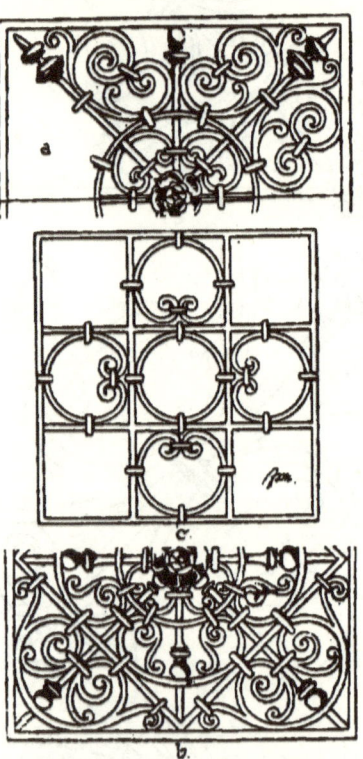

Fig. 121. Modern Grilles.
a. and b. By F. Kiefhaber, Magdeburg.
c. From Venice.

Independent iron gates were not used commonly until the renaissance. This applies as much to the bar as to the geometric designs. Independent rectangular fillings are best adapted for the ornamentation of doors. A frame strong enough to prevent any sagging is filled with the grille work and hung on pivots or hinges and can be made fast by means of buttons, latches, bolts or locks. Where the door is not independent, as for instance when it forms part of a chancel-grille, and where it is incorporated in larger works, the ornamentation of the gate is generally in the same taste as the rest of the design, but made richer in order to distinguish it somewhat from the mass. This applies more particularly to crestings and is especially appropriate where the door is central and dominates the wings (see Fig. 122).

When the door filling presents lines which are not in accord with the general design, they are frequently constructional and are required to counteract the tension and to prevent the door from sagging through its own weight. The line of stress runs diagonally from the lower hinge or angle to the opposite top angle. In double doors the two diagonal ties constitute a symmetrical figure. As this is not the case with single doors symmetry is usually restored by additions that are not actually needed, the simplest taking the form of the St. Andrew's cross. The sagging of heavy doors is most easily prevented by means of a roller fixed beneath the free end of the door,

THE PRINCIPAL FIELDS OF ART SMITHING. 119

and running on a rail. Double doors require stops, unless they close in a complete frame, either a vertical fixed middle bar, or since this

Fig. 122. Grilles in St Ulrich's church, Augsburg. 2nd half of 16th century.

would usually be in the way, an iron stop rising slightly above the ground (the bottom end of a middle bar) or a sill, against which at least the bottom of the door may stop. Where the leaves do not close on each other they cannot be secured by ordinary fastenings (at least not by ordinary locks). These remarks do not apply to renaissance wicket-gates hung in larger doors or gates, to doors of furniture, reliquaries and shrines, &c., because these were of too insignificant a weight to make stops requisite.

Fig. 123. Tabernacle door from the Minster, Villingen.

The bulky locks of the renaissance sometimes produced very disturbing effects in open ironwork grilles, especially when the doors were single. In order to remedy this drawback as much as possible a broad, horizontal connecting-band was frequently introduced at the height of the lock, and such band bore not only the lock, but also a door-knob, or ring. (Compare the two chancel doors, Fig. 123 and 124.)

This cross-band divided the door into two separate panels, just as large doors are often divided into a number of panels (see Fig. 125).

The baroque and rococo periods produced the most important work in connexion with doors and gates. As a rule in these eras works of this kind were created on a far larger scale than in other ages. The doors of churches, palaces and castles, the gates of courtyards and parks furnished occasions for grand and magnificent designs. They mostly consisted of two leaves, each of which was secured to a dressed-stone pillar. The centre is fitted with a broad pilaster-like slam-bar. The transome above the gates is firmly fixed and serves as the closing bar. A rich crown of work is placed above the transome when the gates are free, or a not less rich over-grille is introduced for gates set in an arch. A cross-band, already referred to, is often found as a lock rail at the height of the lock, or else symmetrical ornaments are wrought to the right and left of the slam-bar. Occasionally playful feeling is shown by making the open-work of the gates to represent perspective interiors. The intention in such cases is to create an appearance of space. This leading idea, which is also found in the reduced perspectives introduced into stair-balustrades and entrance halls of the same period is as bold as it is inartistic. The Theresianeum at Vienna and the Constance Cathedral both contain instances of such perspective grille-work.

Doors and gates are also grouped at times by placing single gates at the sides of double ones (see Fig. 127). This idea of a monumental entrance is already seen in the Triumphal Arches at Rome.

Baroque and rococo gates and doors are still found existing in large numbers in the places where they were originally erected, namely in castles and

Fig. 124. Chancel door from the Minster, Thann (Alsace). 16th cent. (Industrial Hall.)

Fig. 125. Gate to the Silver Chapel, Franciscan Church, Innsbruck.

Fig. 126. Gate, 1751. Art-Industry Museum, Leipzig.

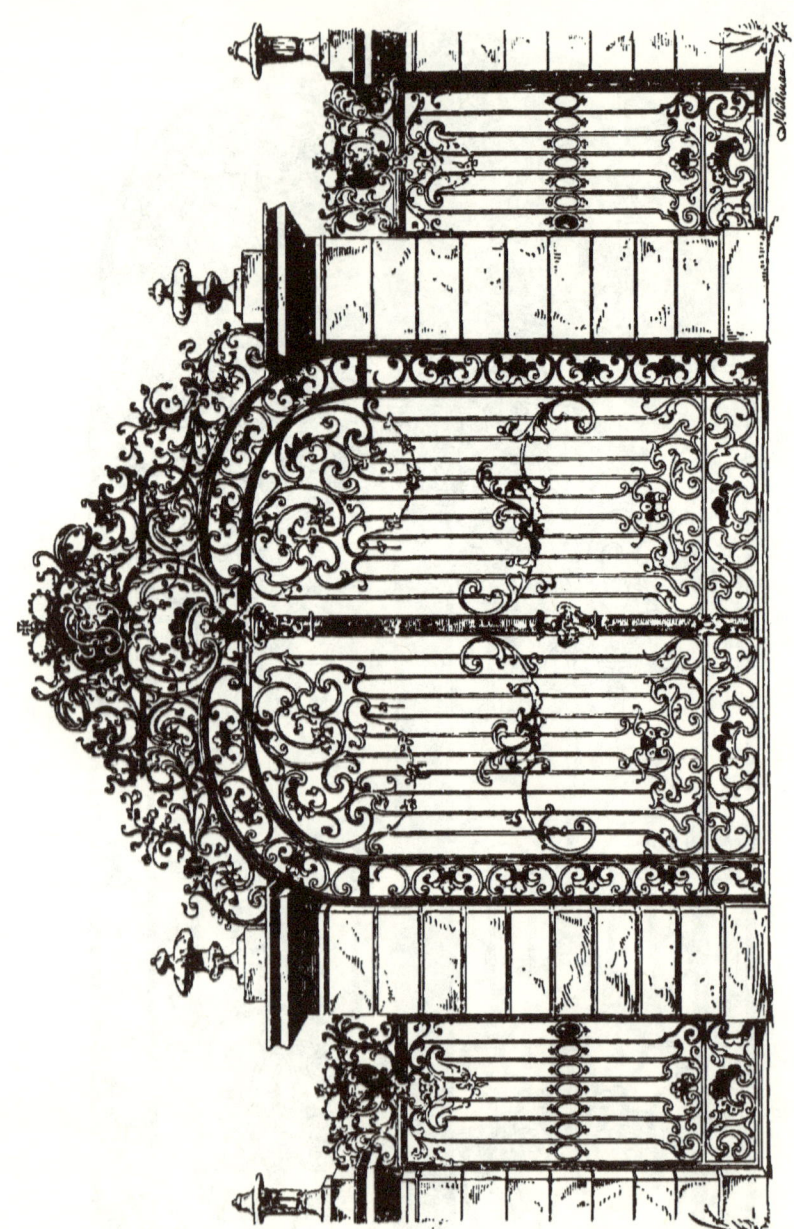

Fig. 127. Park-gates, Castle-Gardens, Karlsruhe.

THE PRINCIPAL FIELDS OF ART SMITHING.   125

parks, palaces and churches in and about Vienna, Munich, Dresden, Würzburg, Schwetzingen, Karlsruhe, &c.

The illustrations of these two periods are limited to one example for the baroque and one for the rococo (see Fig. 126 and 127).

Fig. 128. Modern gates, designed by Director C. Schick, Cassel.

Our modern gates, as seen in garden-entrances, cemetery-railings, &c., are, generally speaking, modest and unimportant in character. The general design of the railing is usually repeated at the gateway where it is often made somewhat richer in execution and strengthened with

126   SECTION IV.

stays. It is only quite recently that one occasionally finds richer ornamental gates in private and public buildings in the larger cities.

Fig. 129.   Modern gate, designed by Peter Sipf, Frankfort on the Maine.

These are in part imitations of earlier periods and partly executed in the modern style of wrought-iron work. Fig. 128 and 129 give examples of such modern gates.

THE PRINCIPAL FIELDS OF ART SMITHING. 127

At Fig. 77 is represented a specimen of a gate top, and three other examples which are available either for gates or for balustrades, are represented in Fig. 130, 131 and 132.

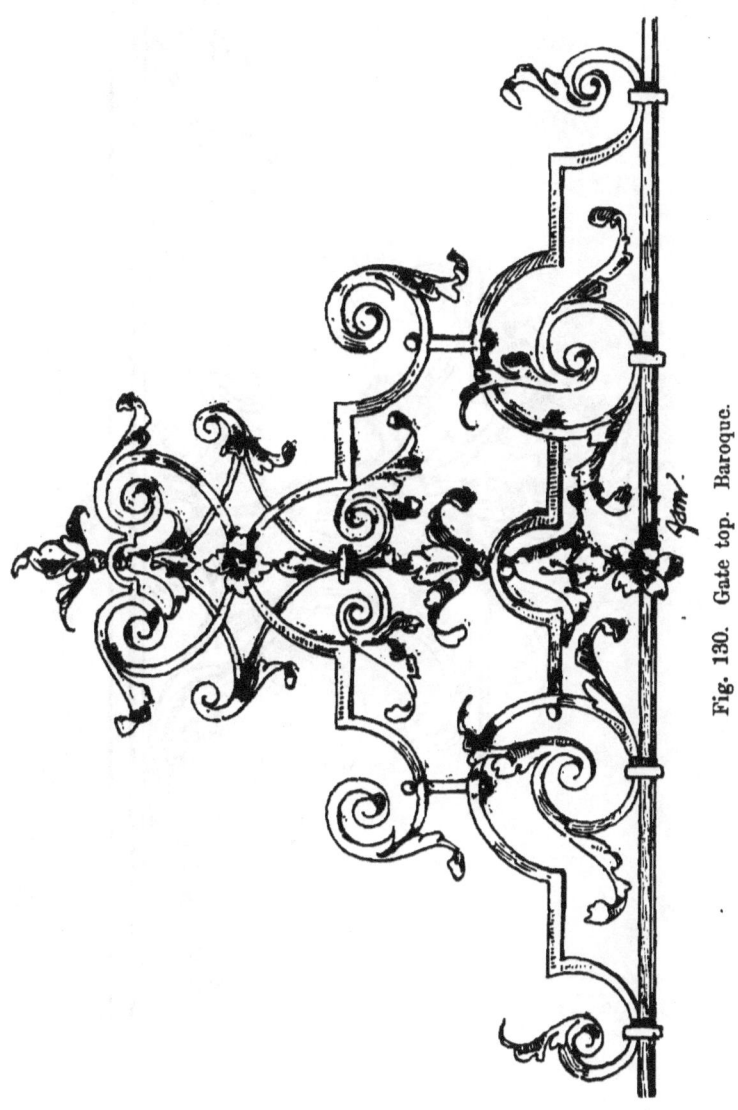

Fig. 130. Gate top. Baroque.

128   SECTION IV.

3. MOUNTINGS.

Wrought-iron mountings are principally used for doors, windows and furniture. These were used most lavishly in the middle-ages and

[Fig. 131. Gate top. Baroque.

Fig. 132. Cresting of grille.

the renaissance period so that, on the average, somewhere about one tenth part of the wood work was covered. From that time mountings became gradually reduced in quantity; they continued to lose in importance until by the time the rococo was reached only about $1/250$th, and later a still smaller fraction of the wooden surface was covered by them. Moreover, brass and bronze entered into formidable competition with wrought-iron for mountings. It is only in the last decade that wrought-iron has resumed its post of honour in this respect, although it cannot be pretended that the demand for it, as yet, stands on anything like an equal footing with that of either the mediæval or the renaissance eras. From what has been said it is evident that our attention must be mainly directed to such former periods.

Directing our minds in the first place to Door furniture we come across various kinds of Strap-hinges, Door-rings, and Door-knockers, as well as Locks. As the next chapter is devoted to the last-named, only the former will be treated of here.

Wooden doors were made out of grooved and tongued narrow boards during the Romanesque and Gothic periods. The hinges generally covered the whole surface of the door; these served partly to bind the wood-work firmly together and partly to form the connexion with the pivots on which the doors turned. These last-mentioned hinges generally stretched across the door and are called Strap hinges owing to their long, narrow form. If another band of iron is run perpendicularly over the other and either rivetted or screwed on at the crossing-point, frequently embellished, with a rosette the Cross-band is produced. Bands to strengthen the door-corners at right angles are called Angle-bands, &c.

These hinges and mounts served two purposes; both to strengthen the door, and to ornament it; the more necessary since the wood-work presented in itself but little scope for embellishment. Doors were consequently sometimes entirely covered with open-work iron-plates, good effects being produced in such cases by judiciously placing the nails, by clever spacing and by embossing or *appliqués*. These effects were enhanced, particularly when painted or heightened with a background, such as cloth or leather.

As, during the renaissance, flat grooved and tongued wood-work gave place to mortised, or in other words, the renaissance doors consisted of a wooden frame-work filled with panels, hinges could no longer cover the whole surface of the door. This was also no longer necessary from the decorative point of view, since the panels were made ornamental by marqueterie or inlay. The development of the hinge had to be confined to the narrow frame-work and became broader than long. The Butt or Hinge, as this form is called, was often ornamented by fretting and lining, by carving and

Fig. 133. Door hinges.

chiselling, embossing from the back or chasing particular parts, engraving, etching, &c. The butts or the pivots were also frequently

Fig. 134. Door hinges, Liège Cathedral. 13th century.

ornamented as they became of more importance owing to the narrowing of the wings or straps of the hinge.

A discussion of the details of form and style may justly be omitted as these have already been referred to in Section III. Fig. 133

THE PRINCIPAL FIELDS OF ART SMITHING. 133

brings together a number of hinges and parts of hinges which suffice as illustrations of the subject. (Thus, for instance, in Fig. 133,

Fig. 135. Wrought-iron door-knockers.
a. Augsburg, 15th century. b. Berlin Museum. c. Munich National-Museum, 16th century. d. 17th century.

7 and 8 represent ends of strap hinges, 9 a "cross-band" hinge, 10, 11 and 12 are butt hinges; also showing the ornamentation of

Fig. 136. Renaissance door-knocker. Fig. 137. Renaissance door-knocker. Fig. 138. Renaissance door-knocker.

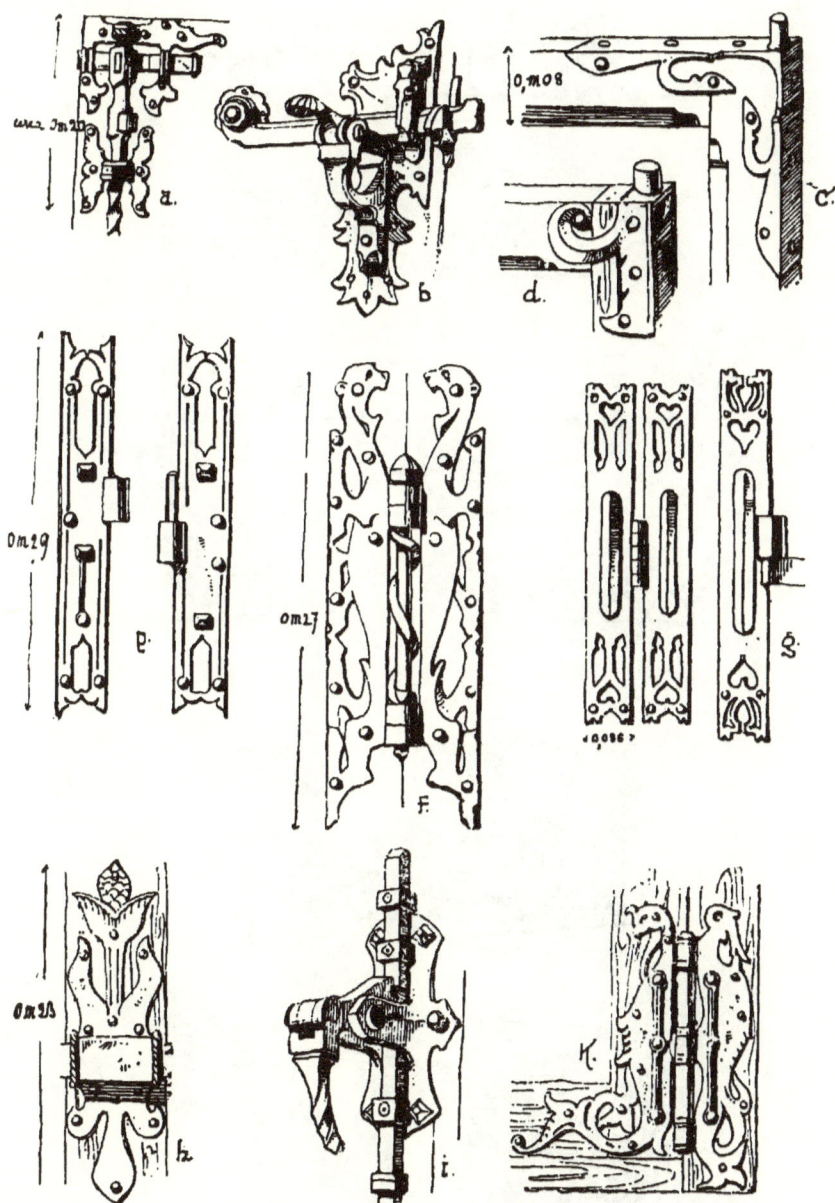

Fig. 139. Window fastenings of the middle-ages, after Viollet-le-Duc.

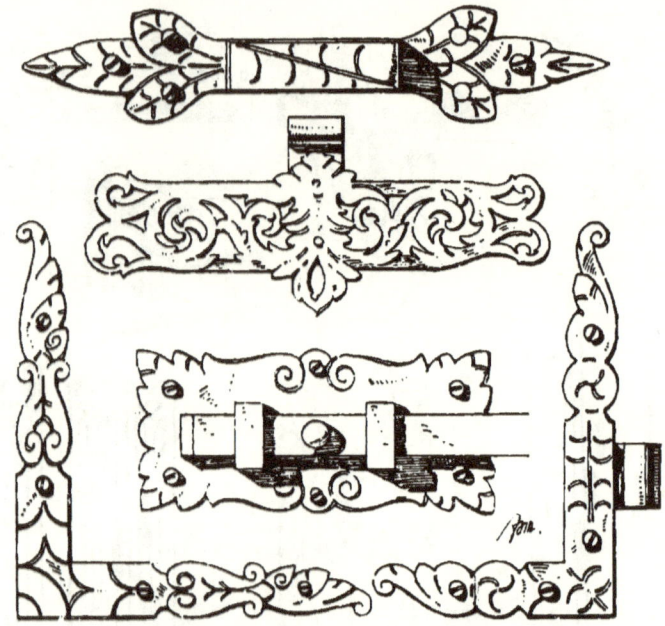

Fig. 140. Renaissance hinges &c.

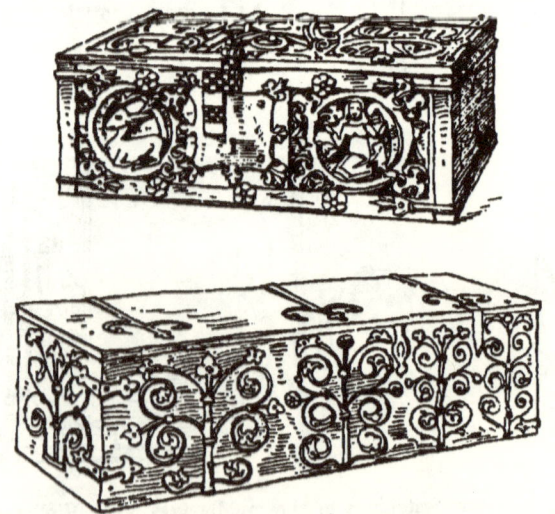

Fig. 141. Chests, of the middle-ages, after Viollet-le-Duc.

THE PRINCIPAL FIELDS OF ART SMITHING. 137

the pivot and its support. Again Fig. 134 represents a very handsome piece of door-furniture at Liège cathedral, of 13th century work and belonging to the transition between Romanesque and Gothic. Door-knockers and -rings were other ornaments almost imperative in the middle-ages and the renaissance. The hammer of the knocker and the closing handle were often one, but sometimes they were separate. Door-knockers were already in use in ancient times, as is evidenced by a specimen Medusa-head and ring found at Capua; the periods of greatest use having been the Romanesque, Gothic and renaissance. In the present day it has almost become obsolete and matter of history, owing to bell-pulls and other methods of ringing. Three different types of knockers can be distinguished. One takes the form of a ring or loop held by a rosette, a lion's head, &c. In this form it also serves as door-handle or closing ring and is still in use for the latter purpose. (Compare Fig. 47 and 135.)

The second type takes the form of a hinged hammer, which is more or less ornamented (see Fig. 137).

The third type is practically a loop drawn out into somewhat the shape of a lyre and in which snakes, human masks and other embellishments not infrequently play a part (see Fig. 136).

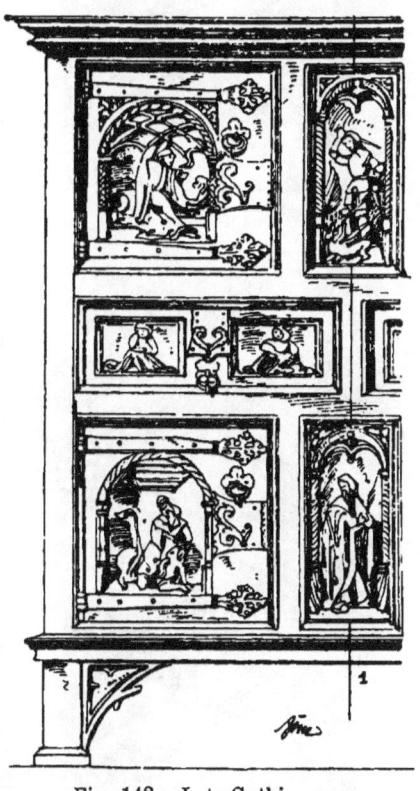

Fig. 142. Late-Gothic press in the Germanic Museum, Nuremberg.

But, inasmuch as these figured and modelled door-knockers make too much demand upon the smith's powers, they are often made of bronze, as is evidenced by many specimens of the Italian renaissance. In all three kinds the moveable portion strikes on a metal protuberance or knob, by which means the necessary noise is produced. In the third group the plate affixed to the door is generally treated as of minor importance, whereas such plate is, in the other instance,

138   SECTION IV.

Fig. 143. Ornament in the Art-Industry Museum, Carlsruhe.

often made to take the first place and is therefore turned into elegant open-work with a back-ground.

As regards Window fastenings the interest is centered more in the mechanical execution than in artistic finish. The wooden window-frame is too narrow to allow of an unhindered spreading-out of the ornament. Butt- and other hinges were therefore used in order to secure the windows to the frame; angled-strips strengthened the corners, while turn-buckles, bolts, latches, catches, espagnolettes and other contrivances served for opening and shutting the windows. Fig. 139 shows a few specimens of mediæval window furniture from Viollet-le-Duc; while Fig. 140 represents others of the renaissance.

Fittings were much used in furniture. In addition to the hinges, locks, looped- and knob-handles, other objects were added that served no practical purpose but that of ornament. Presses and chests were especially subjects for rich ornamentation, and, in these again, the greatest wealth of decoration is found in the middle-ages and the transition to the renaissance. The style of decoration is shown in Fig. 141 and 142 which represent respectively two chests and a linen press.

These ornaments, especially in the later period, were frequently tinned in order to give a more brilliant appearance and to preserve them from rust. In the baroque and rococo periods the hinges, lock-plates, the plates beneath the handles, &c. were cut out of thin sheets and richly embossed. They assumed the character of stamped work, which certainly produced an opulent, sumptuous character, but did not equal the solidity and beauty of the more substantial works of the earlier period (see Fig. 143).

4. LOCKS AND KEYS.

That not only the Greeks and Romans, but also the ancient Egyptians understood the use of locks, is evidenced by among other things, the discovery of keys and of parts of locks. These mechanisms, however, were relatively simple and very different from those now in use. As they only have an archæological interest those who desire to look more closely into the question are referred to special works on the subject, among which Ernst Nötling's "Studies concerning old Roman door and trunk-locks" (published by J. Schneider, Mannheim 1870) must be specially named.

The mediæval and renaissance periods were very inventive with regard to cunning and complicated lock-construction, though the trouble devoted thereto did not always coincide with the degree of security actually attained. Certain it is that the present day produces far greater results with much simpler means. But, on the other hand, these earlier periods showed artistic effects with locks and keys which, generally speaking, we seek in vain at the present day. The main feature has, in modern times, been transferred from the artistic to the essentially practical side.

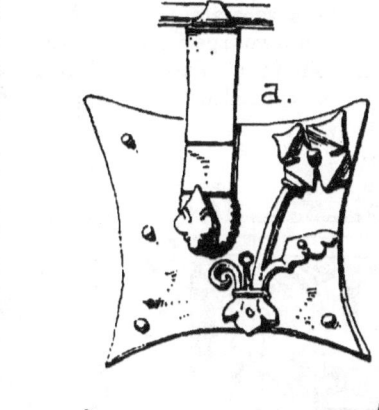
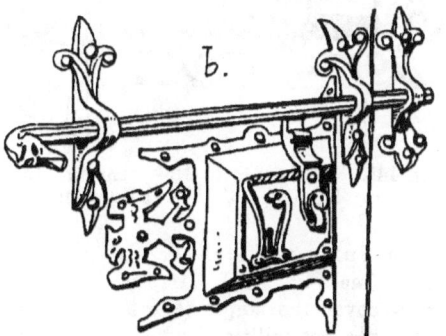

Fig. 144. Mediæval locks.
a. Beginning of 15th century. At Sigmaringen.
b. 13th century. (After Viollet-le-Duc.)

It would lead us too far were all the different constructions of earlier locks to be described. A few remarks must suffice. Even in early times a sort of padlock was in use, which had a moveable hasp, which hooked into the trunk-lock and into which the bolt or catch was shot. Sometimes the method was reversed, so that one end of the hasp was attached to a bolt, while the other entered the lock and was secured (see Fig. 144, *b*). In the latter form it could also be used as a door-lock while the ordinary form was only suited to chests and trunks. The external change of form to which this kind of lock was subjected by suppressing the hasp and causing the bolt to shoot into a separate bolt or box Staple can be recognised by comparing the two specimens shown in Fig. 145. Locks made in the manner shown at *b* are called Box- or Case-locks, because the mechanism is concealed in a case; this distinguishes them from the

Fig. 145. a. Chest-lock, 15th century. In the Bavarian National Museum, Munich. b. 13th century-lock, Angers (Viollet-le-Duc).

Open Spring, Catch or Snap-locks, the work of which is exposed (see Fig. 146). If the lock-case is enlarged in such manner as to cover the staple (and thus prevent the bolt from being pushed back, the possibility of which is also done away with by a box staple) a complete rim lock is the result. The more ancient locks close by means of a so-called Spring or Shooting-bolt, *i. e.* the bolt is pressed forward by a spring and the turning of the key is necessary to unbolt the lock and to open the door or lid. These locks are called German locks in order to distinguish them from the French Tumbler lock (invented by Freitag at Gera in 1724), the kind now in universal use for house-doors. A complete door-lock is made up of a latch, Bolt and a Night-latch. A bolt is called dormant when it is shot out of the catch by a turn of the lever; it is called

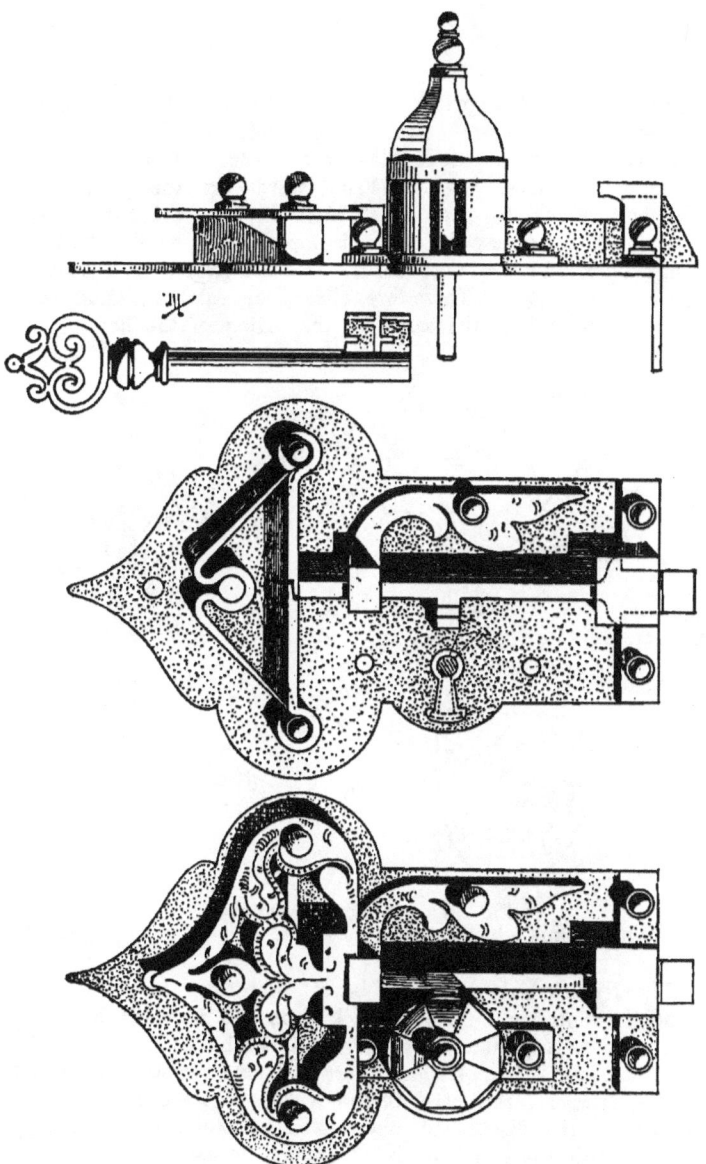

Fig. 146. Old German snap-lock.

142   SECTION IV.

a Spring-bolt when it is pushed backwards and forwards horizontally by the pressure of a spring, like an ordinary bolt. The motion is given by a latch, or bar, lever, or knob-handle.

A bolt is single- or double-turned according as to whether it has one or two-front wards. (Front-wards are the places cut out of the bolt which the key must catch into in order to effect the leverage.) A lock may be made to open from both sides or from one side only. In order to secure the bolt when it is shot home Tumblers are used. These are checks or catches fitting into the bolt and lifted out of the same by the action of the key.

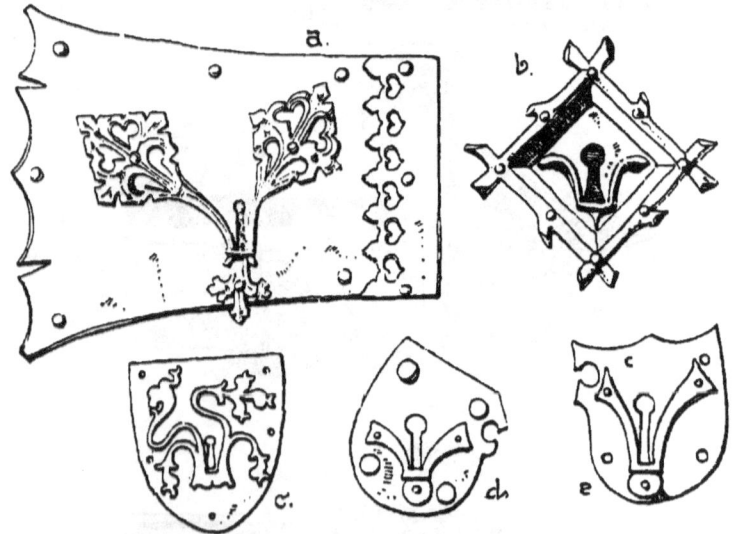

Fig. 147. Mediæval escutcheons.
a. From Cologne. b. From Prague. c. to e. 14th and 15th century.

The Night-bolt can only be set in motion from one side. A protruding knob can be pushed backwards and forwards by the hand, or the motion is produced by turning a knob &c.,

As modern locks, of which there are many varieties (such as combination-, trigger-, Brahma-, Chubb-locks, &c.), are generally devoid of ornament the discussion may be confined to the external parts connected with the construction of ancient locks.

The next point to consider is that affecting Key-plates or Escutcheons. These conceal the places where the woodwork is cut away in order to make room for the key and they also in certain cases (sunk locks) serve for the lever-pin of the bolt; they are often made the subjects of much ornamentation (see Fig. 147 and 148).

Besides the cartouche and foliage it is not uncommon to find figure and grotesque designs, as, for instance, the forms of armoured knights, &c., serving as key-plates.

Almost all the latches in the older door-locks took the form of a sort of door-handle; they were also often ornamented, although such embellishments did not, as a rule, make their use more easy, so that the modern, plain, unornamented door-handles are certainly far more

Fig. 148. Renaissance escutcheons.
a. and b. In the collection of antiquities at Stuttgart. c. In the National Museum, Munich.

convenient. Fig. 149 shows an ornamented door-handle of the time of the renaissance.

The Lock-case is, in its simplest form, prismatically quadrangular and is made up of the lock-plate and sides. These last-named form the Rim and the one in the front, through which the bolt is shot is called the Front-stile. The case may also be made in the form of a very low truncated pyramid as is shown in the locks illustrated by Fig. 145. In richer lock-work the rectangular case is replaced by curved forms (see Fig. 150). This figure shows at the

144 SECTION IV.

Fig. 149. Door-handle and key-plate at the Industrial Hall, Wertheim on the Maine.

same time pretty clearly how these lock plates were ornamented. Etching, engraving, part gilding and also open-work were resorted to.

In Open locks the visible parts of the works are the ones which are often ornamented. The same applies to Puzzle-locks and Escutcheon locks, secret escutcheons being often added in order that only the initiated may be able to insert the key. Fig. 151 represents a modern, open lock constructed in the style of old German locks, the mechanism being shown in the illustration. (Compare this with Fig. 146.)

The Key itself has still to be mentioned. This, so far at least as ornamental specimens are concerned, consists of 4 parts, viz. the Bow, the Boss, the Barrel and the Bit (see Fig. 152). The Bow, *i. e.* the handle, sometimes

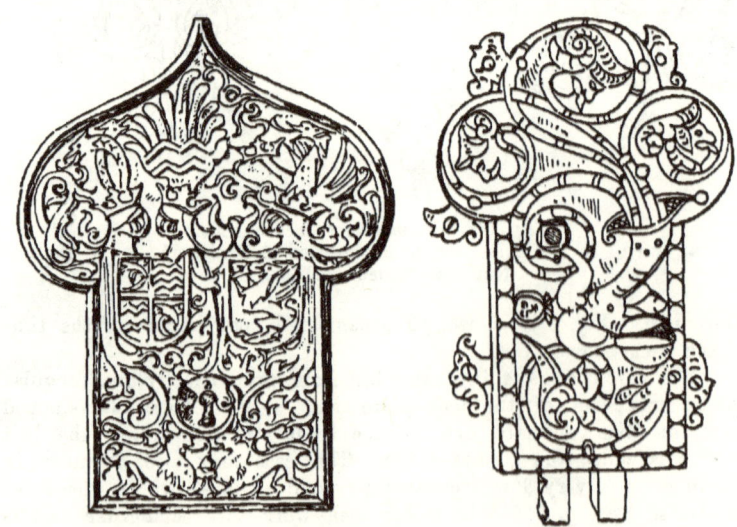

Fig. 150. Ornamental cases to old German locks.

called the Ring, is annular in the plainest specimens, but it may take all sorts of shapes, so that it is found as open-work rosettes, as monograms, as figure work, and even in architectural shapes. The material is not always iron only, as it is common to find brass or bronze bows fixed to wrought-iron barrels. The Boss is

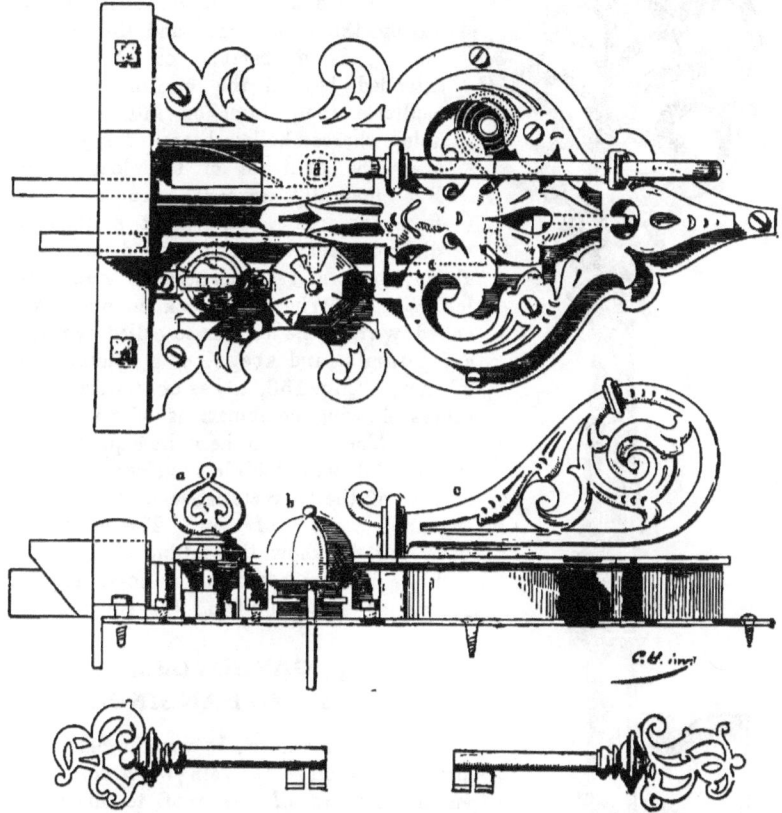

Fig. 151. Open door-lock designed by Director Hammer of Nur

the moulded neck forming the junction between the bow and barrel. It may be very plain or even be dispensed with entirely, but it may also be richly ornamented (see Fig. 152 and 153, c). The Barrel retains its name whether hollow or solid. Keys with hollow barrels are called German or female keys; those which are solid are known as French or male ones. Hollow barrels are mostly round in section, but specimens with triangular or quadrate or stellate sections, &c. are also found.

Cylindrically hollow keys are described as bored; if the section of the barrel shows any other form they are called fancy. Such barrels, which are not easy to make, were much in vogue during the middle-ages and the renaissance. Solid barrels or stems are extended beyond the bit in order to facilitate their insertion into the key-hole, whereas hollow-barrelled keys pass over the drill-pin of the lock. The Bit generally appears quadrate (seen from the front), whereas it may show different shapes when looked at from above, bearing resemblances to numerals, letters, &c. Where the lock countains so-called wards the bit has so-called wards, and guards, and steps, or a combination of these. (Fig. 153, *a* has only wards, the others showing combinations of wards and steps.) Those shown here and previously in Fig. 50 will doubtless suffice to give an idea of the characteristics of keys from the artistic point of view. The enormous keys used to show the calling of the guild and the beakers made in key-form may be mentioned before quitting the subject.

5. GARGOYLES, BRACKETS AND HANGING SIGNS.

As is well known, it was formerly the practice to allow the rain-water collecting in the gutters of the roof to discharge itself through gargoyles directly into the street instead of carrying it down to the ground by means of pipes, as is now the case. These gargoyles projected over the street. It is less the aim to discuss these water-spouts which, taking the forms of dolphins, masks, &c. were mostly made of sheet-iron, than to discuss the wrought-iron bearers and stays which supported them. These bearers were mostly plain, rod-like props or stays, but often they were richly ornamented,

Fig. 152. Key, 17th century.

the lines of stress commonly governing those of the ornament: Fig. 154 shows such an ornamentation.

Fig. 153. Renaissance keys.

Wrought-iron brackets were used for the most varied of purposes. They generally served as bearers, the object to be supported being either placed upon the free end or hung from it, the other end of the bracket being fixed to the wall in such wise that it was either rigid or else worked on pivots.

In the mediæval age large brackets were often put up in the churches to remove the covers of baptismal fonts. These covers were either raised by means of a chain running through a pulley, or else moved aside by means of a swinging bracket. A portion of one of these brackets has already been shown in Fig. 49.

At this same period brackets already served as candle-sticks (see Fig. 155). The idea of bracket-lights retains its influence to the present day. Fig. 156 presents an example in the Italian late-Renaissance style, while Fig. 157 shows us a handsome modern gas-bracket.

Fig. 154. Details of a stay German, late Renaissance.

148   SECTION IV.

The bracket was also frequently used in early times as the bearer of **guild-** and **trade-signs**. Locksmiths were particularly given to show their calling by hanging out a key, as may be seen in Fig. 158.

The like idea was adopted for inns and taverns, the signs being sometimes painted and sometimes executed in relief in wrought-iron, according to the nature of the design. Fig. 159 and 160 show such

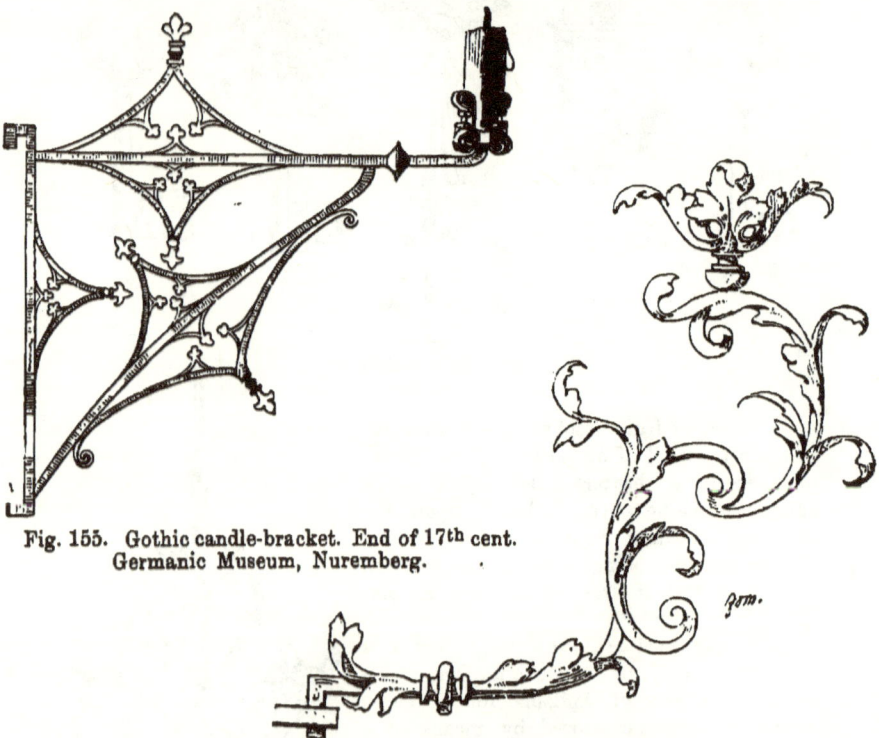

Fig. 155. Gothic candle-bracket. End of 17th cent. Germanic Museum, Nuremberg.

Fig. 156. Gilt Light-bracket. Verona Cathedral.

tavern-signs. Wall-brackets generally take the form of a console or of a right-angled triangle. The first-mentioned form (see Fig. 163) is the more effective from an aesthetic point of view; the last is the more constructive, as the principal rod running out from the wall is supported by a stay set at an angle, thus leaving a triangular space to be filled in with ornamentation (see Fig. 161 and 162).

The main arm is also frequently supported from above, in which case the triangle is formed on the upper side. These supporting stays are ornamented similarly to those of gargoyles.

THE PRINCIPAL FIELDS OF ART SMITHING. 149

Brackets are much used in modern times for supporting sign-boards serving as trade-advertisements, &c. These tablets may be made rectangular, circular, elliptic or any other shapes according to fancy, the frame being suitably embellished. Cartouches with enrolled volutes make satisfactory designs for such signs; moreover, these scrolled ornaments, cut out of

Fig. 157. Modern gas-bracket, designed by Ad. Haas.

sheet-iron, are easily made. Some modern sign-boards are shown in Fig. 164 to 166. Furthermore Fig. 167 illustrates a rich frame for a sign without the bracket.

In order to prevent sign-boards from swaying in the wind they are usually made completely fast to the bracket, even when they appear to hang loose. As it produces an inelegant effect to run the bracket direct and crudely out of the wall a back-plate is generally introduced to which the arms are secured either in form of a flat bar fixed to the wall, or a small cartouche is employed. Where the main iron of the bracket is inserted into the wall a rosette may also serve to

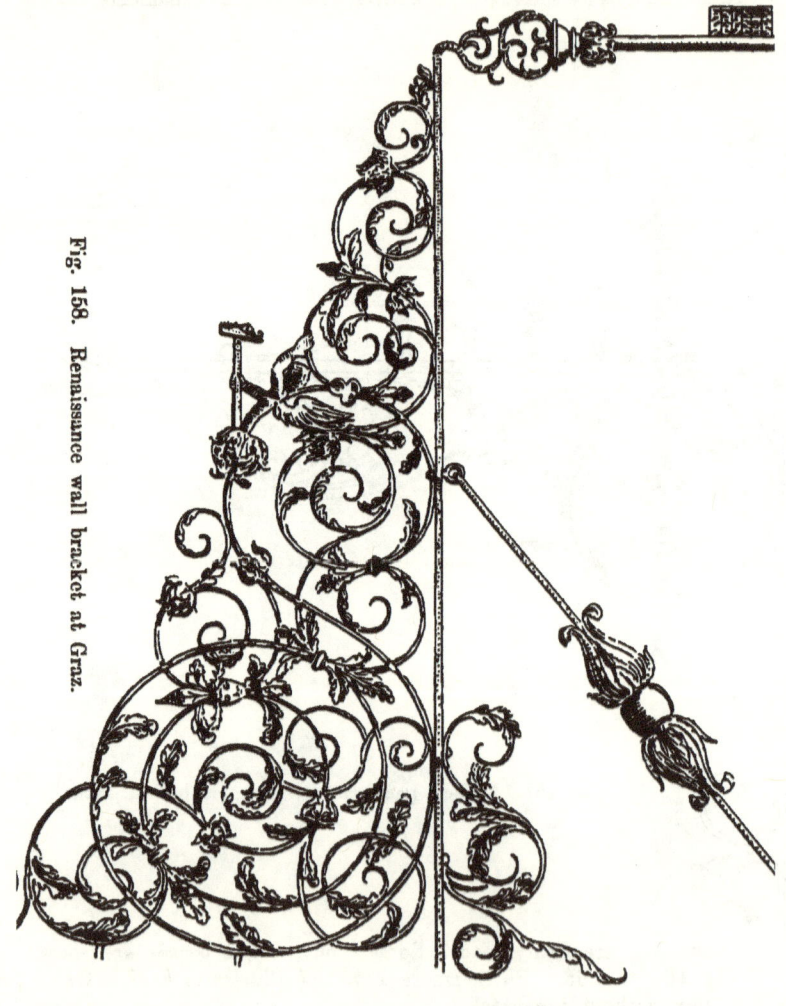

Fig. 158. Renaissance wall bracket at Graz.

remove the impression of bareness. When a bracket is used for gas or electric light a tube is employed in order to supply the gas or carry the wires.

6. CANDELABRAS, CANDLESTICKS, CHANDELIERS, CORONAS AND LANTERNS.

Wrought-iron was employed in very early times as a material for various illuminating-apparatus, because like other metals its in-

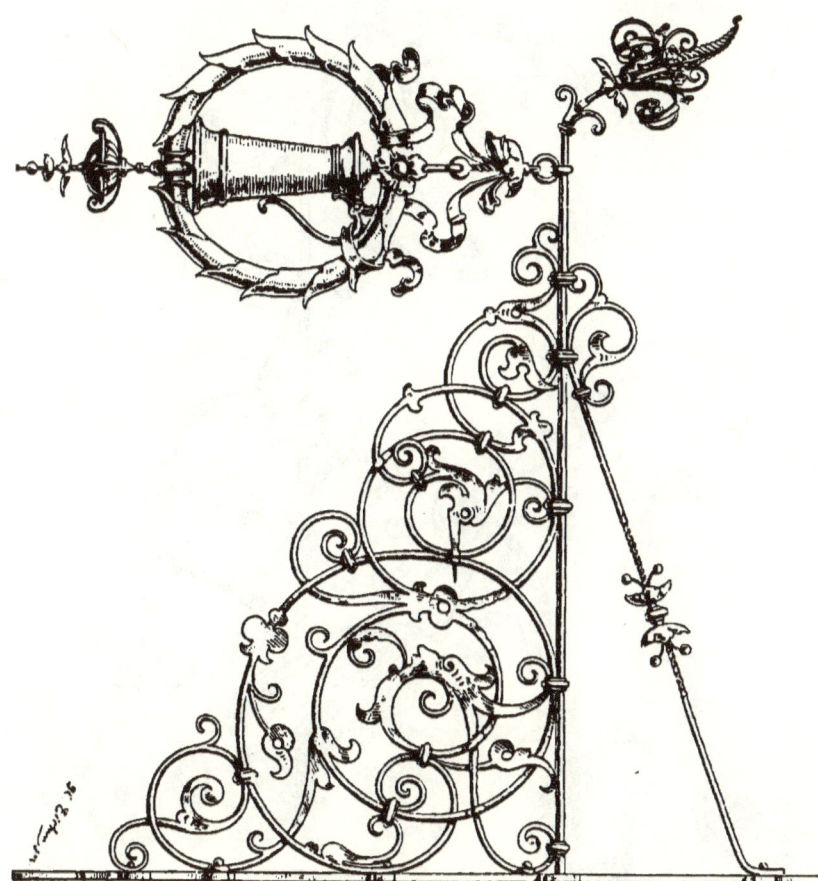

Fig. 159. Modern tavern sign, Renaissance style, designed by Director C. Schick, Cassel.

combustibility makes it specially suited to the purpose. The form and finish of the apparatus varied of course with the progress of time as much as the method of lighting itself changed. Oil- or lamplight, candle- and torchlight, gas- and electric light, each require specific and distinct arrangements. The more ancient appliances for giving light stand in certain contrast to those of the present day,

Fig. 160.
Sign of the "Crocodile".

which arises from the fact that the first often combined high artistic finish with great imperfection from the practical point of view, whereas modern lighting-apparatus immensely surpasses the ancient in respect of technique and utility, but does not, as a rule, equal them in respect of art.

Fig 161. Renaissance-wall-bracket, from Innsbruck.

Fig. 162. Baroque bracket, from Zurich.

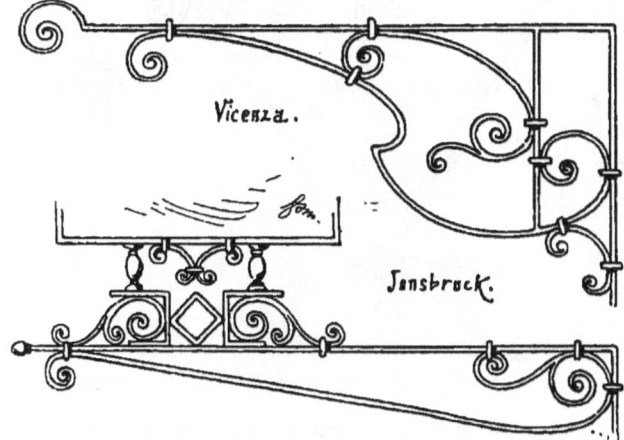

Fig. 163. Bracket, designs from Vicenza and Innsbruck.

If the various lighting-appliances are classified, the following types will appear distinct from each other, namely: the upright candle-stick, which, in larger sizes, is also known as a standard-candelabra, the hand- or portable candlestick, the bracket-

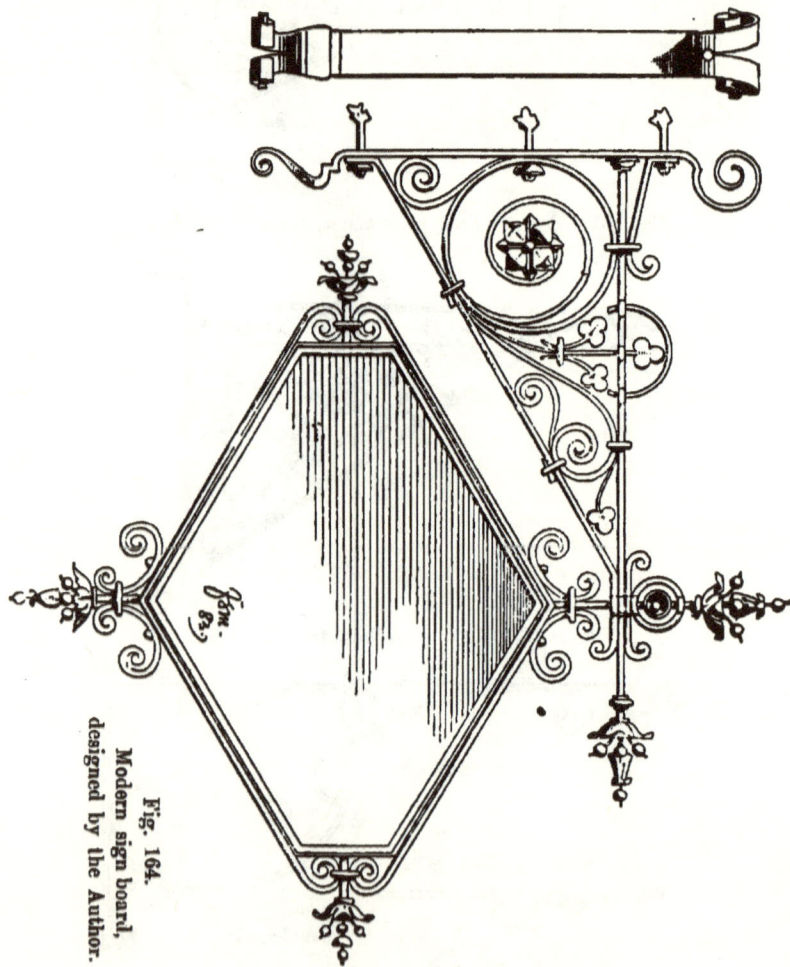

Fig. 164.
Modern sign board,
designed by the Author.

light, the lanterns and hanging or suspended lamps, the coronas and the modern petroleum lamps. Let these be discussed in their order.

The upright or standing candlestick was the candle-holder of

THE PRINCIPAL FIELDS OF ART SMITHING. 155

the mediæval and renaissance periods and is still that of modern times. The word is derived from candela = a candle. The middle-ages gave the preference to conically-tapered prickets; the present age is more inclined towards cylindrical sockets in which the candle

Fig. 165. Modern sign board, designed by Prof. Th. Krauth, Carlsruhe.

is fixed. The candlestick generally consits of a base, a shaft and nozzle. The first is often round and flattish, or else a tripod in the style of the antique candelabra feet, in order to secure greater firmness. The nozzle at the top generally finishes off with a candle pan or plate in order to catch the melted drops. If this saucer is removable it is called a "bobèche". In standing candelabra the upper part spreads out into branches provided with candle-sockets. In mediæval times comparatively high candelabra were not uncommon, especially in

Fig. 166. Modern sign board, designed by E. Crecelius.

churches (see Fig. 168). The renaissance and baroque periods show very handsome and rich specimens, for instance that of the 17th century (Fig. 169). Earlier ages also furnish us with models for

THE PRINCIPAL FIELDS OF ART SMITHING.   157

imitation (see Fig. 170 and 171), the former representing a small modern candlestick, the latter a large upright candelabrum, both from the workshop of E. Puls of Berlin.

Fig. 167. Sign board frame. Designed by the Author.

By Hand-candlestick is understood a portable candlestick frequently provided with a handle. It is of modest size and dwarf,

Fig. 169. Three branched candelabrum, Baroque.

Fig. 168. Candlestick in San Pedro, Tarrasa, Spain, 14th century.

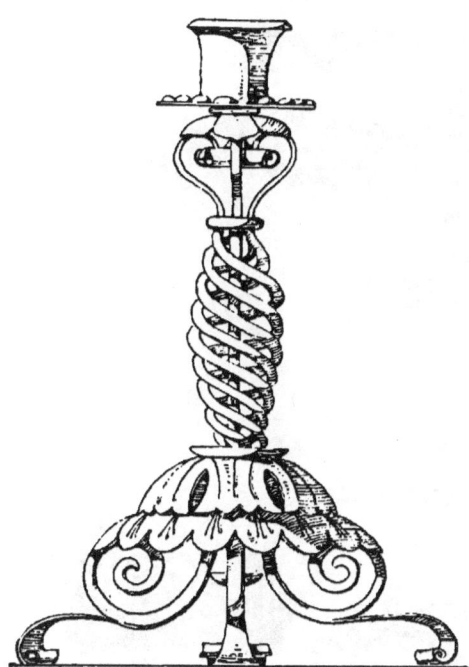

Fig. 170.
Modern candlestick; by E. Puls, Berlin.

Fig. 172.
Renaissance candlestick. 17th century.

Fig. 171. Modern candelabrum, designed by Architect Zaar, executed by E. Puls, Berlin.

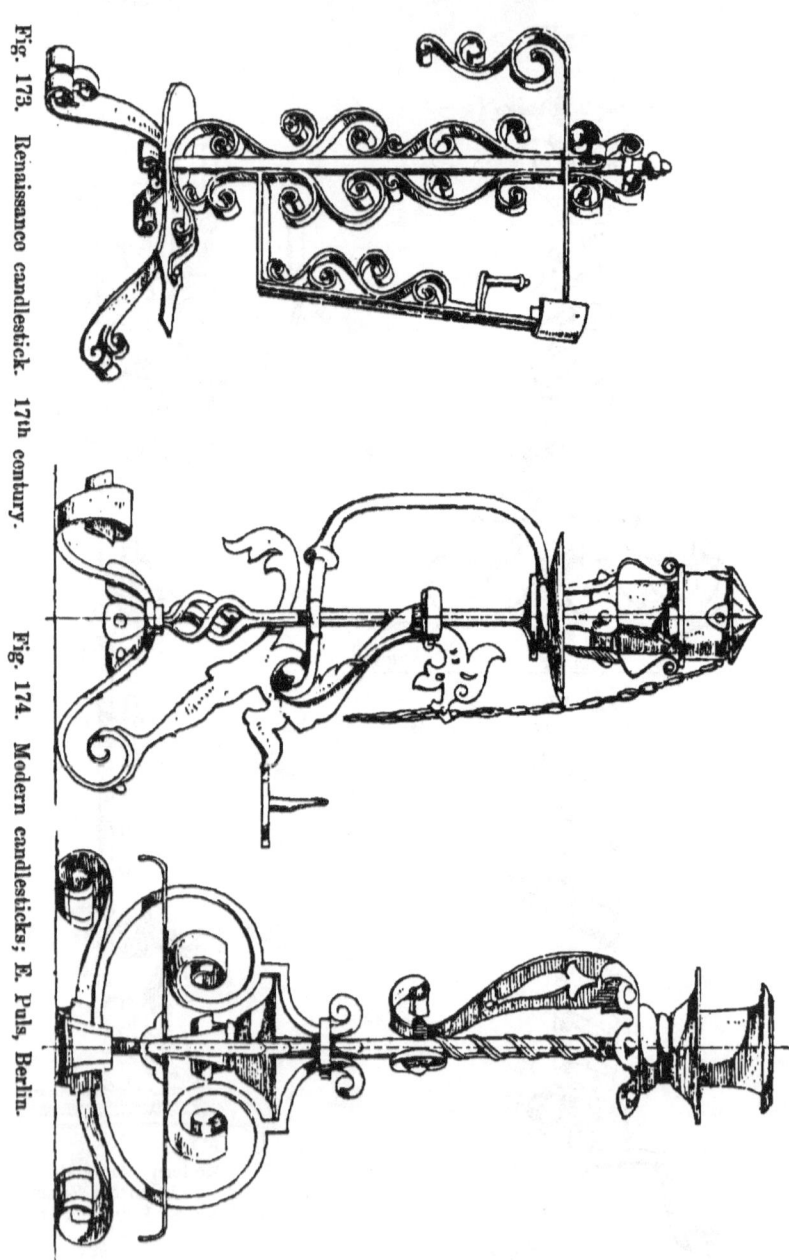

Fig. 173. Renaissance candlestick. 17th century. Fig. 174. Modern candlesticks; E. Puls, Berlin.

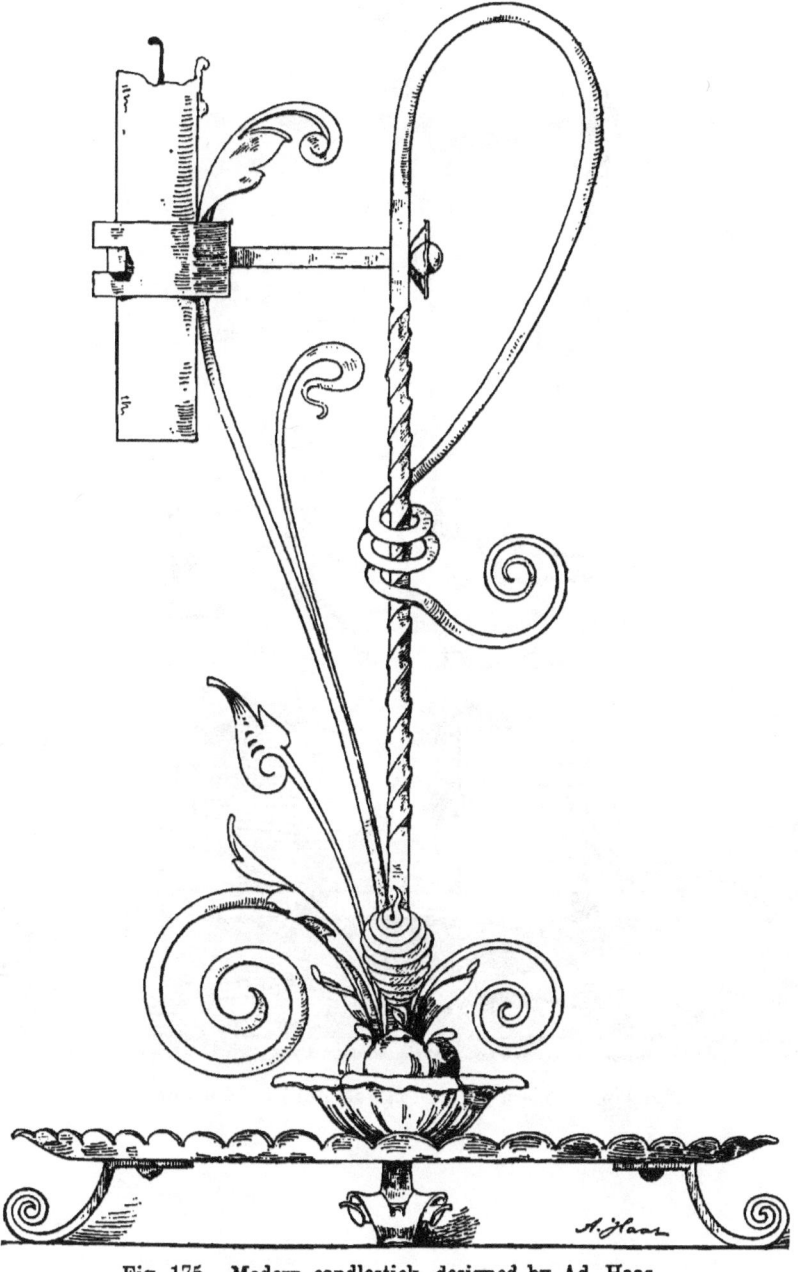

Fig. 175. Modern candlestick, designed by Ad. Haas.

Fig. 176. Modern candlestick, designed by the Author.

generally broader than its height. The renaissance was much given to them and developed a great wealth of designs. Candlesticks in the style of Fig. 172 are frequently met with. The socket holding the candle could be moved spirally up and down, according to the length of the candle. Invention was active at this period in devising adjustable mechanism to such kinds of utensils (see Fig. 173). Sometimes an extinguisher was added, as shown in two of the modern examples shown in Fig. 174 and 176. Wrought-iron portable candlesticks have of late come much into fashion and offer occasion for very original designs.

Wall lights are, as the term implies, lighting appliances which are fixed to walls pilasters, columns, &c. Sometimes they are secured permanently, so as to be immoveable, at other times they are made

Fig. 177. Candle bracket, German Renaissance. National Museum, Munich.

moveable on a pivot or hinge. They may be made to hold one or many lights. In arrangement and finish they are usually similar to the brackets already discussed, the branches being fitted with prickets or sockets, or with the requisite burners, according as to whether they are to serve for candles, oil or gas, &c. Compare Fig. 60, 155, 156 and 157. Fig. 177 shows a further specimen of ordinary renaissance form, while Fig. 178 represents a modern swinging bracket for one candle ornamented with flowers and foliage.

Hanging lamps and lanterns may be open or closed, that is glazed with glass. They are the outcome of the necessity of having a light which could not be upset, which could be raised and lowered and which could be protected against the influence of wind and draughts. Hanging lamps and lanterns are specially suited to open air uses, corridors, vestibules, stair cases and similar spaces. They came into use in early times and are capable of being produced in

164 SECTION IV.

handsome, elegant designs without any difficulty. Owing to the great improvements made in glass manufacture of late years it has become possible to produce ball-shaped and other forms of bent glass and

Fig. 178. Modern candle bracket, designed by the Author.

thus, where so desired, to dispense with the box-like shapes glazed with sheet glass to which former times were confined. Fig. 179 shows an open Gothic lantern made to hold a number of candles, while Fig. 180 depicts a closed modern lantern glazed with roundels.

Coronas arose through the desire to have a number of lights arranged in a circle and suspended. In the middle-ages the lights were preferred on one horizontal plane, as shown in the renaissance example Fig. 181. Later styles preferred them in several tiers and this principle holds good with regard to gas and other chandeliers to the present day. Very pleasing effects are produced when the number of lights in each row is properly proportioned. Candle and oil coronas may he hung with chains running on a pulley so as to permit of their being raised or lowered. In gas coronas the chain is replaced by the pipe which supplies the gas, the necessary movement being effected by means of a cup and balljoint and a stuffingbox. More than 5 or 6 arms are rarely set in the circle; where more lights are desired the arms are subdivided into smaller branches, as is commonly done with the ordinary bracket-lights. Fig. 182 shows two simple modern hanging-lamps in wrought-iron. Fig. 183

Fig. 179. Gothic lantern German work. (Formenschatz.)

and 184 show two other, somewhat richer examples which were shown at the Carlsruhe Art smith's work Exhibition.

The modern Petroleum lamp has also proved an incentive to experiments in wrought-iron, though this appliance does not appear

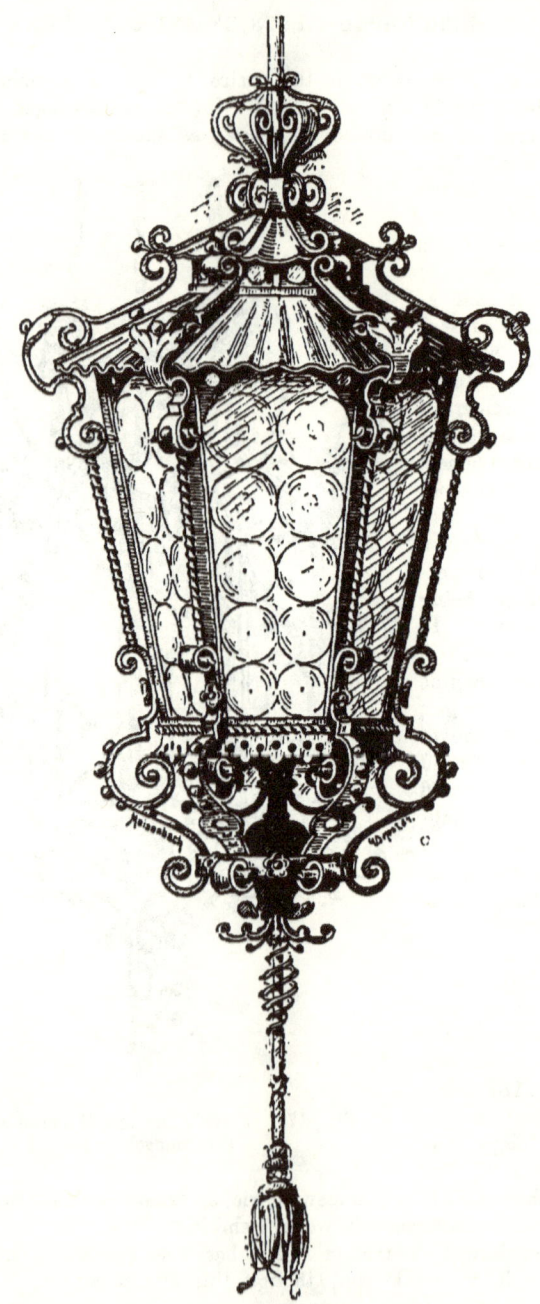

Fig. 180. Modern lantern, by Emil Bopst, Berlin.

to be well suited to iron-work, and has moreover brought with it many offences against style and good taste.

7. WASH STANDS AND FLOWER STANDS.

The renaissance smith had evidently the antique tripod in mind when he first attempted to design the three-legged wash stands.

Fig. 181. Wrought-iron Renaissance corona.

Such articles are often found on Italian, as well as on German soil, and they frequently show great richness combined with excellent workmanship. The frame-work is generally made of square iron, while the ornamental additions are of slighter bar-iron. In combination with this is often found a holder for a water-jug; and even a place for hanging towels. Fig. 185 shows a wrought-iron tripod of Italian origin dating from the 17th century. This is already very

Fig. 132. Modern Gasaliers. (Krauth and Meyer's work on locksmithing.)

Fig. 183. Modern chandelier, by F. Lang, Carlsruhe.

Fig. 184. Modern chandelier, by H. Hammer, Carlsruhe.

Fig. 186. Modern flower stand, designed by H. Grisebach, executed by P. Marcus, Berlin.

Fig. 187. Modern flower stand, designed by the Author.

Fig. 188. Aquarium stand, designed by F. Miltenberger.

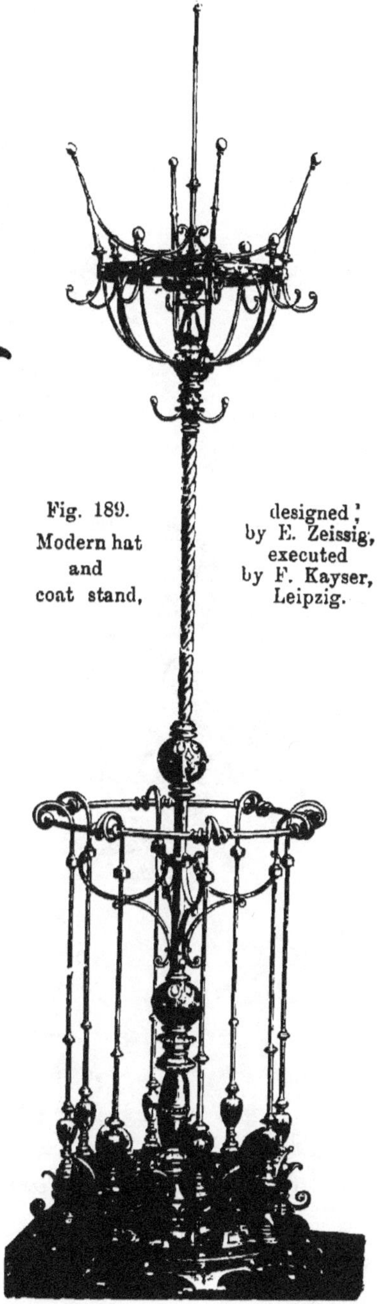

Fig. 189. Modern hat and coat stand, designed by E. Zeissig, executed by F. Kayser, Leipzig.

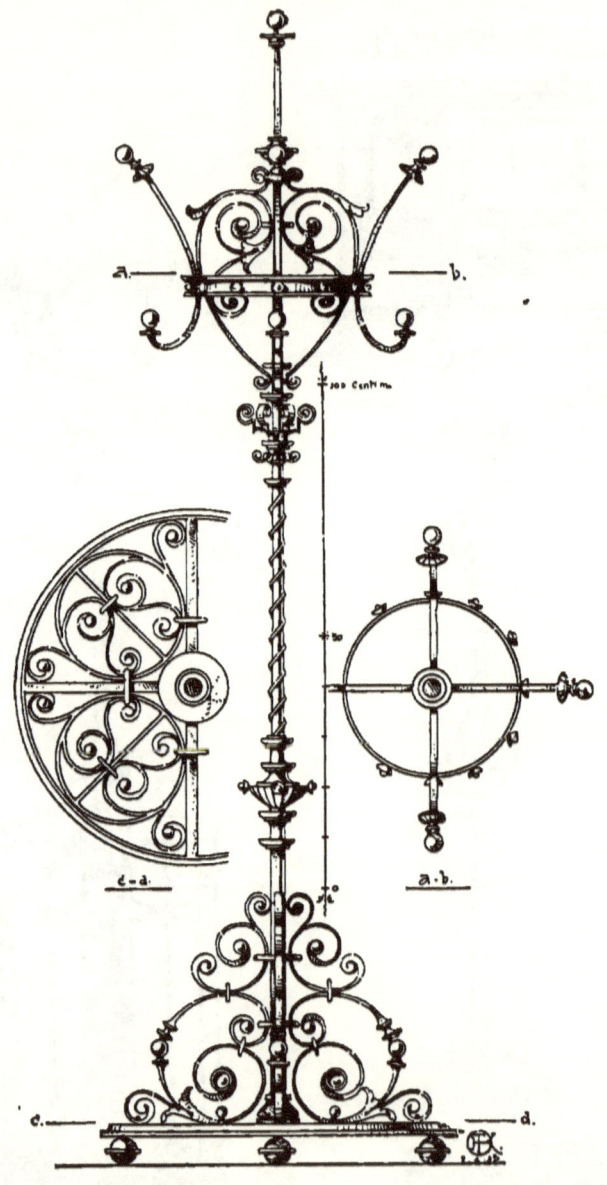

Fig. 190. Modern hat and coat stand, designed by Prof. Th. Krauth.

THE PRINCIPAL FIELDS OF ART SMITHING. 175

rich in style and it must be admitted that more simple designs generally give better outlines.

Such tripods have greatly come into fashion of late years; they

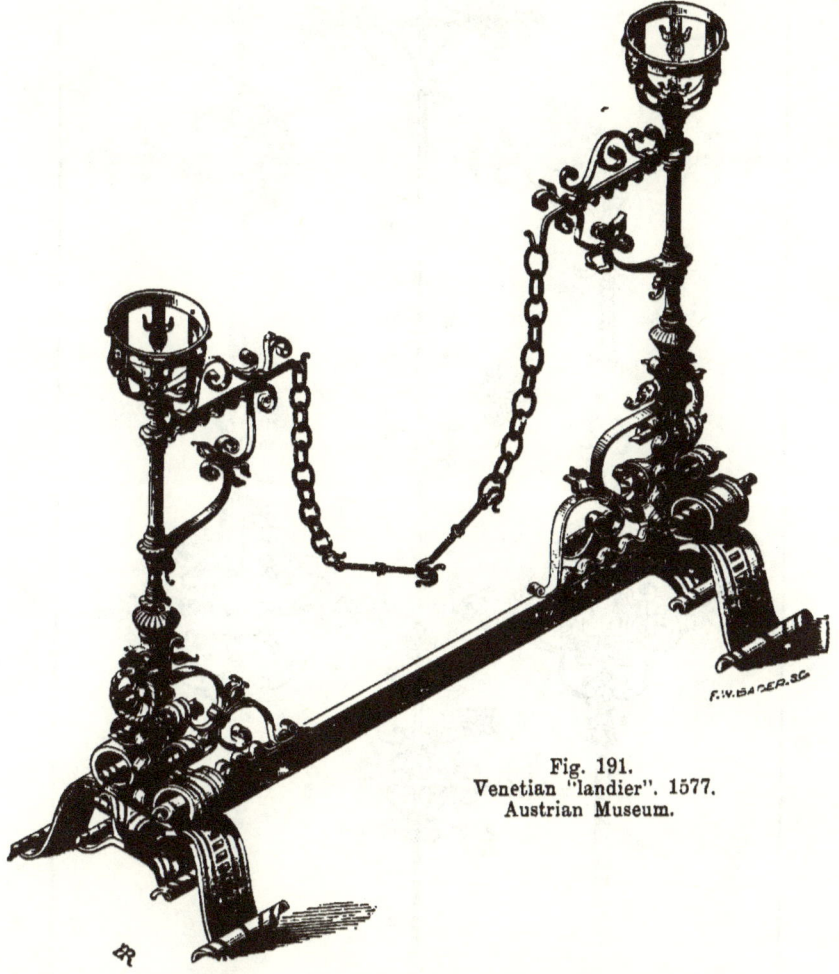

Fig. 191.
Venetian "landier". 1577.
Austrian Museum.

serve as wash stands, as stands for wine-coolers, take the form of flower stands, occasional tables, &c. Painted porcelain plates, rich majolica ware, or metal trays are used with them as card trays, and table-tops. Flower tables, on the other hand, are fitted with

Fig. 192. Various crosses for towers and graves.

THE PRINCIPAL FIELDS OF ART SMITHING. 177

sheet-metal revolving trays, &c. by which, the flower-pots can be turned towards the light without the necessity of moving the whole

Fig. 193. Grave cross. Art-Industry-Museum, Berlin. Renaissance.

stand. Fig. 186 and 187 illustrate two modern flower-stands, one being carried up into a second tier.

Fig. 188 shows a wrought-iron stand for an aquarium; appro-

Meyer, Smithing-art. 12

Fig. 194. Grave cross designed by Prof. Krauth.

priate to its purpose it takes the form of a high table with a small top. Similar forms are employed for reading-desks, only that in such case the top is sloping instead of flat.

Fig. 195. Modern grave cross, designed by the Author.

This is the place to mention the hat and coat-stands which have of late been frequently made of wrought-iron and which often combine a stand for sticks and umbrellas, when the tripod gives place to a sheet-iron tray. Two specimens are shown in Fig. 189 and 190. There are no old models for such furniture and it is improbable that they were known.

Finally, reference may be made to andirons or fire-dogs, which where often made of wrought-iron during the middle-ages, the profile frequently taking the appearance of standards (see Fig. 191). As with the open fire-places, these utensils have fallen into disuse.

8. CROSSES FOR GRAVES AND TOWERS.

In early times, from about the end of the middle-ages, the latin form of cross was made ornamentally of wrought-iron in order to provide finials for the summits of towers, spires and gables of churches and chapels. This symbolic ornament was often very simple, though no less frequently very rich and elegant. The actual construction of the cross was generally of stout bar-iron, while the ornamental additions were of lighter make. The arms of the cross were generally finished off with leaves or flowers, in plainer work with spear-heads or lance-points; the upper arm was also frequently made to serve as a vane or weather-cock. The rectangular spaces between the arms were ornamented with rings or scroll designs, which served at the same time to strengthen the whole. Whereas the crosses on gables were, almost without exception flat, and on one plane, spires, &c. were sometimes fitted with crosses, the arms of which were not only directed to right and left, but to the front and back, radiating from the centre and thus producing a richer perspective effect.

During the renaissance it was also customary to decorate graves with iron crosses. A very great number of these grave-crosses may be found in old German churchyards. The baroque and rococo periods and also that of Louis XVI retained this practice; later on stone monuments took the place of such crosses, and it is only quite of late years that the old custom is being revived. Grave-crosses differ from crosses for spires, &c. in more ample detail, since they may be inspected closely and moreover they bear memorial tablets.

These tablets are often placed inside a metal case for the sake of protection; they frequently contain, in addition to name, dates of birth and death of the deceased, a religious or secular quotation, as, for instance:

> (on the outside)
> "Fear and love for God he ever bore;
> Stranger, now unclose the door!
> (on the inside)
> Here he rests from earthly pain;
> Stranger, close the door again!"

Fig. 196. Grave cross, by P. Marcus, Berlin.

Fig. 197. Grave cross, designed by E. Bopst, Berlin.

Is not this both naive and pious at the same time and a well-planted hint to the inquisitive to close the door of the tablet in order to secure a longer duration to the record of the departed?

Among the illustrations to this chapter Fig. 192 shows a sheet from the author's work on "The science of ornament" on which two old grave crosses are seen with some old and modern spire crosses. Fig. 193 to 197 represent five other grave crosses, one old and four modern, of which the last has, in addition to the memorial-tablet-case, a basin to hold holy water. In certain districts are also found arrangements to receive flowers, candles, &c., according to the customs prevalent in such parts.

9. ARMS AND ARMOUR.

The chapter on arms is of prominent interest even although the very many still extant specimens may no longer directly serve as models for the smith-craft of the present day. The manufacture of modern weapons is quite different, and generally speaking devoted strictly to mechanical accuracy so that the old masterpieces apart from their historical interest, only furnish ornamental studies applicable to other branches of the art.

The armourer's art is of the oldest and was already highly developed in antiquity. Damascus was one of the cities where it was first practised. Damascus-blades have a world-wide reputation that extends back for thousands of years. Their elasticity and toughness was obtained through repeated weldings and blendings of small, thin metal-plates or wires of iron or steel of various degrees of hardness. The patterns brought out on the surface by etching with acids is called "damask grain", and these patterns vary greatly according to the way in which the metal has been worked. The armourer's art passed on from Damascus and from the East in general until it reached the Greeks and Romans, who, however, made use of iron or steel for the blades only, employing other materials for the handles and ornamental additons, and for shields and armour. After the collapse of the universal Roman domination, what remained of the craft of the antique armourer in the confused migration of races is survived to be developed in the western lands of the middle-ages.

In the time of Charlemagne, armour helmets and shields were already made of iron or else strengthened with iron. From the 11th to the 14th centuries warriors wore iron shirts of mail; the full suits, of plate armour, came into use later on. It is such armour that presents the highest interest as to its technique and artistic decoration, for it comprises the finest work ever wrought in iron, or, it may be said, which skilled labour has ever produced. A new industry, one strictly restricted within recognised limits, was soon

Fig. 198. Old sketch design for a decorated suit of armour.

THE PRINCIPAL FIELDS OF ART SMITHING. 185

developed, which again became divided into distinct branches and special guilds. The cities of Augsburg, Nuremberg, Munich, Landshut, Milan, and many others, produced an imposing number of very

Fig. 199. Ornamental details of a helmet; 17th century.

important metal-beaters, harness-makers, helmet-makers, &c. The great collections of armoury such as those of the Armeria at Madrid and of Turin, the Ambras collection at Vienna, that of the Bargello

Fig. 200. Ornamental detail of the breastplate of a suit of Christian II.

at Florence, &c., show us a wealth of most wonderful work, the major part of which emanated from German armourers.

The helmet, the gorget, the breast- and backplates, the shoulder guards, the arm-guards, the thigh- and leg-guards, the gauntlets and sollerets, with many others were all necessary parts of a complete suit of armour (see Fig. 198).

When one considers how difficult it sometimes is to obtain a well-fitting suit of clothes from the modern skilled tailor, it will

Fig. 201. Sword-handle. Bargello, Florence.

be easy to imagine what difficulties were connected with the production of a suit of armour of satisfactory fit. To this difficulty was superadded the ornament required by rich and distinguished personages for parade and other important occasions. To these requirements are due the very pearls of art. Besides the hollowing out and embossing which of themselves demanded unusual skill and which, moreover, served in part as ornamentation, there was the etching, engraving, inlaying and gilding.

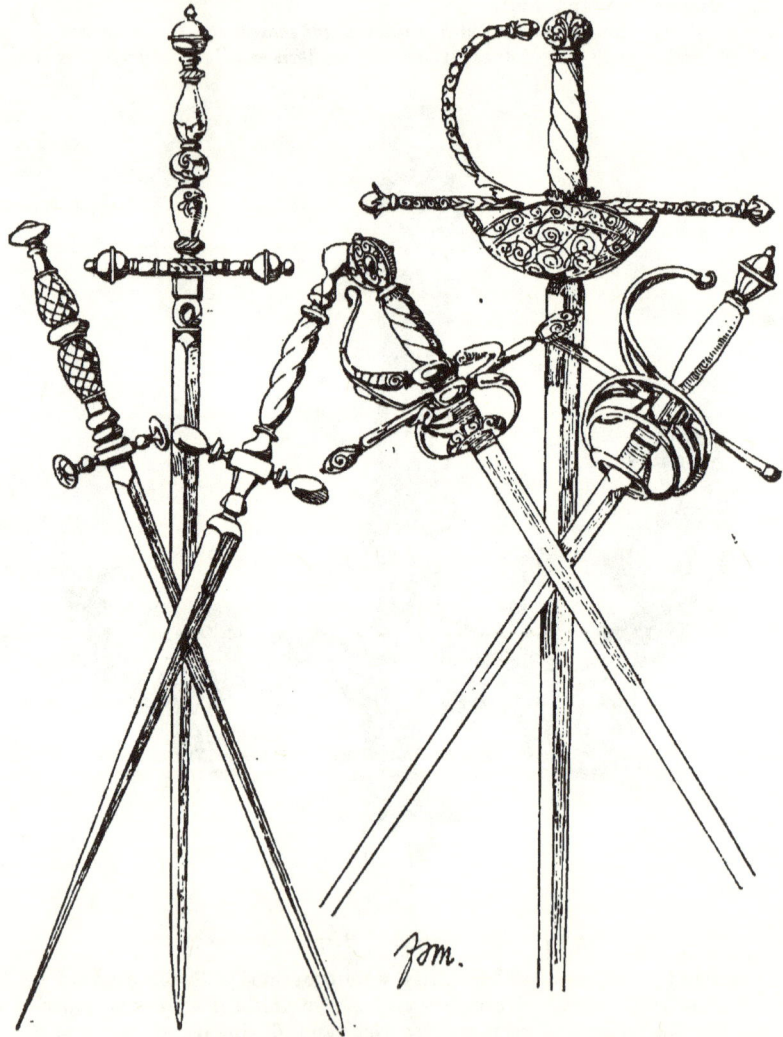

Fig. 202. Various swords and daggers.

One is simply astounded at the correctnees of style and the excellent effects which resulted from ornament produced with the aid of these

Fig. 203. Partisans, halberds, pikes, bills, fauchards, &c.

Fig. 204. Pikes, halberds and partisans. Bargello, Florence.

Fig. 205. Halberd. 16th century.

processes. It must however be remembered that such first-class artists as Dürer, Holbein, Miehlich, Aldegrever, and Burgkmayr did not find it beneath their dignity to furnish armourers and sword-cutlers with sketches and designs for their work. Fig. 199 shows sundry ornamental details of an etched helmet, and in Fig. 200 is seen the ornament on the breast plate of a German suit of armour.

Helmet and armour are defensive weapons to which the shield must be added.

In the early part of the middle-ages this was not made of iron, owing to its large dimensions, so that it was, at most, only studded with iron. Later on, in the renaissance, it became gradually handier and smaller until at last, as with armour, it was dispensed with altogether, or only used for show. Iron became the favoured material for shields, and their surface was decorated in harmony with the rest of the armour. The forms which occur most frequently in parade-shields of the later period, are the circular, those with more or less fanciful cartouche shapes, and the almond shape.

The helmet was next in importance to the shield and received the most conspicuous attention and honour. Whereas the old forms of heaume, the flat and round helmets, the helm, basinet and armet were, comparatively speaking, plain and simple, the Burgundian helmets, "bourginots" and morions, which came up later, often showed an over-rich ornamentation.

Among the offensive weapons the one- and two-handed-swords were the most important and general. However this cut-and-thrust weapon may vary in respect of size and finish, three parts are invariably found. These are: 1) the blade, which may be sharp on one or both sides, and at the point which is generally more or less tapered; it is mostly straight and less often curved (like a cutlass), rarely waved, and sometimes, in order to reduce its weight, it has so-called blood-grooves; it is mostly plain or ornamented with etching or engraving. 2) the hilt or handle which holds the tang of the blade and is fitted with either a pommel, a cross, a sword-shell or basket. 3) the sheath or scabbard with or without belt. It is self-evident that the hilt and scabbard were the parts most ornamented, and to this end other materials were often used. Here, again, it is to the ornament rather than to the practical use that attention has to be given. The hilts deserve more especial notice; it will be seen that many of these are made of iron and that very considerable pains were taken in piercing and working the specimens which are to be found in all collections of importance. (Compare Fig. 201 and 202.)

The dagger or poignard is a sword in miniature in which the cross-part may either disappear altogether or assume a smaller form. Sometimes it is furnished with a guard, but never with a basket-hilt. Among the ornamentations of the scabbard the so-called "dance of death" is a favorite and frequent subject.

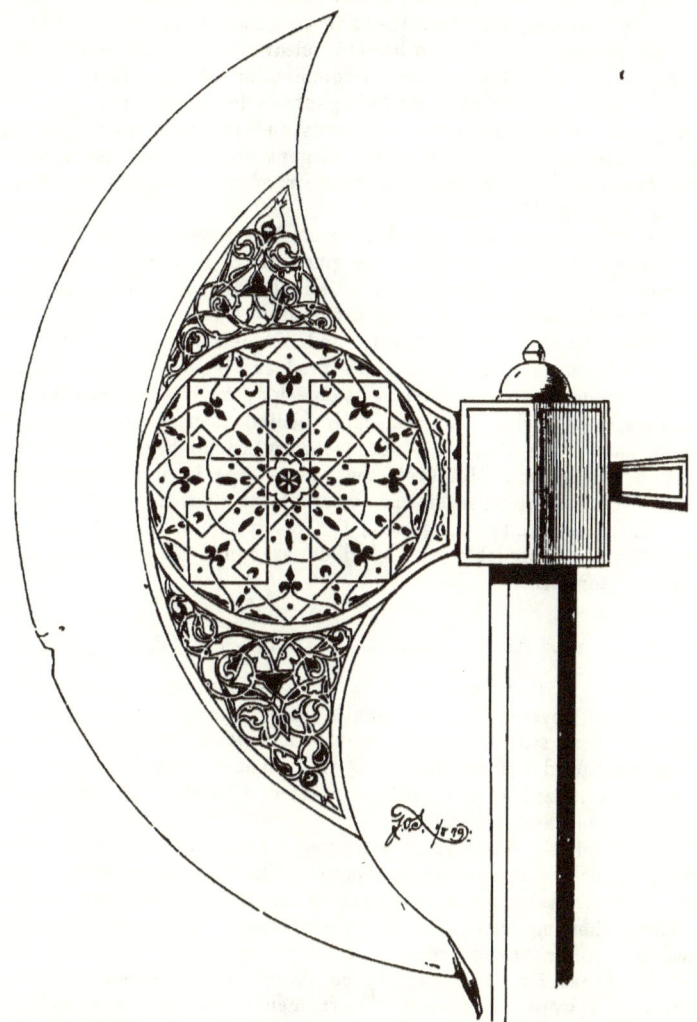

Fig. 206. Gold inlaid Hispano-Moresque battle-axe. Bargello, Florence.

To the mediæval and renaissance weapons of offence, which take many forms, belong furthermore: **spears and lances**, which consist of a wooden shaft to which is fitted an iron spike shaped like a leaf or an awl; partisans, which are spears with flat blades and symmetrical side-points; bills, fauchards and forks, taking the forms of sickles, scythes, &c.; battle-axes and hammers; and halberds (derived perhaps from "helmbarte", cleave helmet, or from "halbe barte", a half-blade) which was a combination of the pike or partisan with the battle-axe; clubs, morning-stars (clubs with iron prickles), flails, and many other arms. Halberds and partisans more especially call for attention, owing to their elegant outlines and the splendid ornamentation of their blades. A number of such weapons are shown in Fig. 203 to 206.

If we compare the flint, bone and fish-spine weapons of primitive races with those which are the glory of the renaissance, we obtain a survey of the two extremes of artistic handicraft in a special field, most of which are inconceivably different.

The invention of gunpowder and the introduction of fire-arms brought this special art to an undeserved end. Guns have undeniably afforded opportunities for artistic decoration, as is evidenced by our museums, nevertheless they are so far behind their precursors both in number and kind that it is not necessary to go beyond the simple mention of the fact. Moreover, the present age is, unfortunately, too matter-of-fact and practical to care to ornament its weapons of sport and war; where such attempt is made the result is not always happy, in spite of the innumerable and glorious examples handed down to us from our forefathers.

10. ALL OTHER OBJECTS IN IRON.

Although it was attempted in the previous pages to arrange the whole field covered by the smith into separate and appropriate chapters, there still remain a number of objects not dealt with, and to be mentioned before concluding this manual. Some of these were formerly produced by the guilds connected with this art, while others are still made and used.

First of all there are the horse muzzles which, to a certain extent, belong to the preceding chapter. These peculiar objects, which have gone entirely out of use, are to be found in collections, some of them showing such splendid workmanship that they deserve, at least, not to be overlooked.

Bellcots for small bells in courtyards and passages, as well as bell-pulls were often made of ornamental wrought-iron and such are coming into vogue again (see Fig. 207 and 208). The last, especially when they take the form of natural floral hangings, seldom fail in producing a good effect.

Vanes or weathercocks made of pierced sheet-metal with

THE PRINCIPAL FIELDS OF ART SMITHING. 195

elegant outlines were articles in moderate request from early times and are continually required at the present day (see Fig. 209).

Vessels, such as lamps and field flasks are now and again found of iron; the latter of large dimensions, holding as much as 12 gallons, and even more (see Fig. 210).

Table necessaries, at least knives and forks are often made of iron, down to the present time, and this not only as regards the blades, but also the handles.

The same remark applies to scissors and shears. Of course the ornament is chiefly confined to the handles; for when the blades are embellished it can only be by means of engraving, etching or inlaying (see Fig. 211).

Every day working-tools or such as serve for special purposes, such as hammers, tongs, fire-irons, compasses, &c., are also found in museums when ornamented, as in the Germanic Museum at Nuremberg. These, like many other things connected with the art, present in themselves little more than historical value however interesting they may be. Many of these articles were the results produced by journeymen or masters as proofs of their skill and title to be received into the locksmiths' or smiths' guild, and entailed a sacrifice of time and pains in no way commensurate with the price generally paid for such work.

Fig. 207. Bellcot. Upper Austria.

Among the articles now made of wrought-iron which are at present popular, and which, possibly, may become still more so, are caskets; etching and applied pierced open sheet-metal form the ornament, the handles and lock giving scope for further embellishment. In like manner, and rightly so, the frames of fire-screens are of late made of wrought-iron,

13*

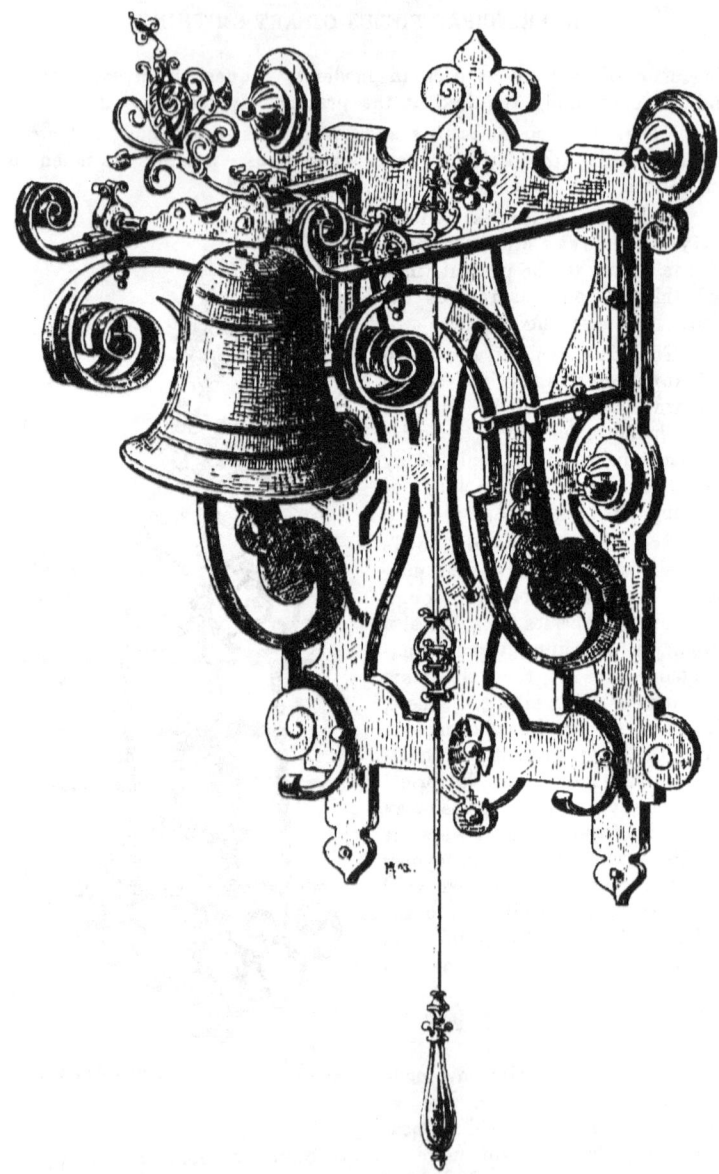

Fig. 208. House bell, designed by Arwed Rossbach, executed by Herm. Kayser, Leipzig.

THE PRINCIPAL FIELDS OF ART SMITHING. 197

while the actual screen consists of stuff, leather, &c. embroidered and painted, &c. (see Fig. 212).

The present age also aims at making writing and smoking utensils, card cases, picture and mirror frames, clock-cases and table-ornaments of wrought-iron (see Fig. 213).

These are risky experiments, mere concessions to fashion, and best left alone. Such things may be attempted in Delta-metal, to which the same technique applies, and when made of it they look

Fig. 209. Wrought-iron Weathercock.

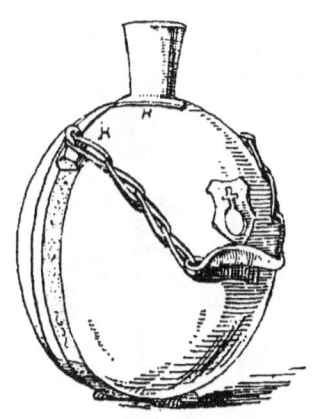

Fig. 210. Wrought-iron field-flask (Viollet-le-Duc).

more distinguished and suited to the purposes for which they are intended, have a better colour and do not rust.

A few words concerning rust — this foe to wrought-iron — and the means of making it innocuous will not be out of place and may serve to close the fourth section of this manual. Inasmuch as it is not a pleasurable task to have to polish and grease articles every two days, they are often covered with a coat of colourless lacquer. But if this is to be thoroughly effective it must be laid on

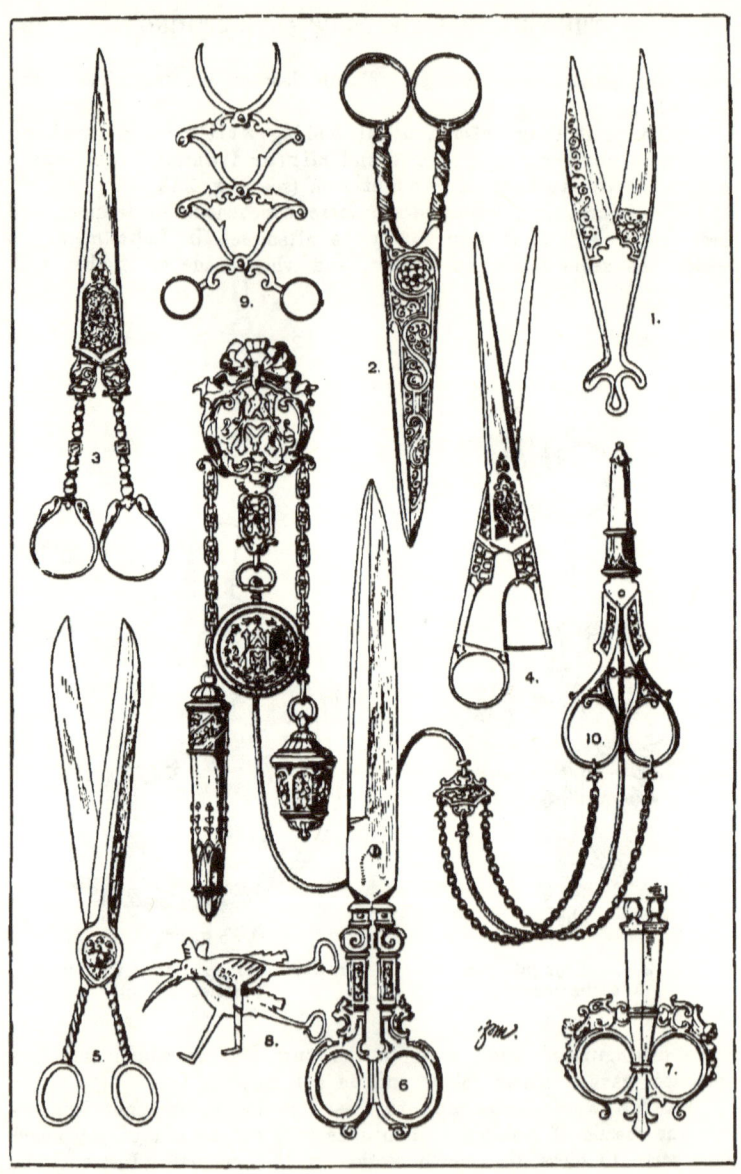

Fig. 211. Scissors.

Fig. 212. Modern fire-screen by P. Markus, Berlin.

Fig. 213. Wrought-iron clock-case by Reinhold Kirsch, Munich.

THE PRINCIPAL FIELDS OF ART SMITHING. 201

thickly, and in consequence the lustre thus produced detracts from the appearance of the object. It is true that tinning, nickeling, or gilding, thoroughly protects from rust, but what becomes of the character of wrought-iron? Moreover, to nickel or gild articles all over produces an unrestful and disagreeable effect. There remains the method of tempering with oil in the fire. This is about the best mode of treatment, but only on condition that it is properly carried out, so that it does not form a sticky, dirt-collecting and dirt-producing surface. Furthermore, there is the coating with oil-colour, which course is not to be despised when the articles are of any considerable size and when it is done with skill and judgment in appropriate style. Polychromatic treatment, which was formerly frequent, although not, generally speaking, executed in particularly good taste, seems to have attracted too little attention at the present time. A few, well-softened tones and free from gloss will always produce an agreeable effect. This proceeding certainly demands a considerable artistic sense of harmony, and this every locksmith and every house-painter cannot be expected to possess. But, let the attempt be made, and if not successful the first time, perhaps repeated attempts may lead to a satisfactory result. Practice is often better than theory. Both must, however, go hand in hand if any good is to result.

And this is the aim of modern art-industry.

May it prove successful!

Fig. 214. Details of the ornament of a shield.

SUPPLEMENT.

TABLES OF WEIGHTS AND MEASURES.

a. German sheet-iron scale.

Scale N°	Thickness in millimeters*)	Weight per square meter in kilog.**)	Scale N°	Thickness in millimeters	Weight per square meter in kilog.
1	5,50	44	14	1,75	14
2	5,00	40	15	1,50	12
3	4,50	36	16	1,375	11
4	4,25	34	17	1,250	10
5	4,00	32	18	1,125	9
6	3,75	30	19	1,000	8
7	3,50	28	20	0,875	7
8	3,25	26	21	0,750	6
9	3,00	24	22	0,625	5
10	2,75	22	23	0,5625	4,5
11	2,50	20	24	0,5000	4,0
12	2,25	18	25	0,4375	3,5
13	2,00	16	26	0,3750	3,0

b. German wire-scale (millimeter-scale).

This scale comprises 100 N^{os}. To find the diameter divide the particular N° by 10. The following table gives the N^{os} at intervals of 5, thus 5, 10, 15, &c.

Scale N°	Diameter in millimeters	Weight per linear meter in grammes	Scale N°	Diameter in millimeters	Weight per linear meter in grammes
5	0,5	1,5	55	6,5	181,5
10	1,0	6,0	60	6,0	216,0
15	1,5	13,5	65	6,5	253,5
20	2,0	24,0	70	7,0	294,0
25	2,5	37,5	75	7,5	337,5
30	3,0	54,0	80	8,0	384,0
35	3,5	73,5	85	8,5	433,5
40	4,0	96,0	90	9,0	486,0
45	4,5	121,5	95	9,5	541,5
50	5,0	150,0	100	10,0	600,0

*) An English yard is mtr. 0.91439 or $914^{39}/_{100}$ millimeters. The meter has 1000 millimeters, or 100 centimeters.

**) An English pound (avdp.) weighs $453^{59265}/_{1000000}$ grammes. The kilogramme has 1000 grammes.

c. *Table shewing the dimensions and weights of wrought-iron gas-barrel.*

Inside measure (diameter) in millimeters	Inside measure (diameter) in English inches	Thickness of metal in millimeters	Outside diameter in millimeters	Weight per meter in grammes
6,5	1/4	3,25	13	600
10,0	3/8	3,5	17	850
13,0	1/2	3,5	20	1150
16,0	5/8	3,5	23	1550
19,0	3/4	3,5	26	1750
25,5	1	4,0	33,5	2500
32,0	1 1/4	4,5	41	3400
38,0	1 1/2	4,5	47	4300
51,0	2	4,5	60	6000
57,0	2 1/4	6,0	69	8300
63,5	2 1/2	6,25	76	9000
76,0	3	6,5	89	11500

d. *Table of weights of round bar iron.*

Diam. in mm	Weight per m in kilos	Diam. in mm	Weight per m in kilos	Diam. in mm	Weight per m in kilos	Diam. in mm	Weight per m in kilos
5	0,153	21	2,70	44	11,86	76	35,38
6	0,221	22	2,97	46	12,96	78	37,27
7	0,300	23	3,24	48	14,12	80	39,21
8	0,392	24	3,53	50	15,32	85	44,26
9	0,496	25	3,83	52	16,57	90	49,62
10	0,613	26	4,14	54	17,86	95	55,29
11	0,741	27	4,47	56	19,21	100	61,26
12	0,882	28	4,80	58	20,61	105	67,54
13	1,035	29	5,15	60	22,05	110	74,12
14	1,201	30	5,51	62	23,55	115	81,02
15	1,378	32	6,27	64	25,09	120	88,21
16	1,568	34	7,08	66	26,69	125	95,72
17	1,770	36	7,94	68	28,33	130	103,53
18	1,986	38	8,85	70	30,02	135	111,65
19	2,213	40	9,80	72	31,76	140	120,07
20	2,452	42	10,81	74	33,55	150	137,84

e. *Table of weights for square bar iron.*

Thickness in mm	Weight per m in kilos	Thickness in mm	Weight per m in kilos	Thickness in mm	Weight per m in kilos	Thickness in mm	Weight per m in kilos
5	0,195	21	3,44	44	15,10	76	45,05
6	0,281	22	3,78	46	16,51	78	47,46
7	0,382	23	4,13	48	17,97	80	49,92
8	0,500	24	4,49	50	19,50	85	56,36
9	0,632	25	4,88	52	21,09	90	63,18
10	0,780	26	5,27	54	22,75	95	70,40
11	0,944	27	5,69	56	24,46	100	78,00
12	1,124	28	6,12	58	26,24	105	86,00
13	1,318	29	6,56	60	28,10	110	94,38
14	1,529	30	7,02	62	29,98	115	103,15
15	1,755	32	7,99	64	31,95	120	112,32
16	1,997	34	9,02	66	33,98	125	121,88
17	2,255	36	10,11	68	36,07	130	131,82
18	2,526	38	11,26	70	38,22	135	142,16
19	2,816	40	12,48	72	40,44	140	152,88
20	3,120	42	13,76	74	42,71	150	175,50

f. Table of weights for flat bar iron

Thickness in millimeters

Breadth in millim.	3	4	5	6	7	8	9	10	11	12
10	0,234	0,312	0,390	0,468	0,546	0,624	0,702	0,780	0,858	0,936
12	0,281	0,378	0,468	0,562	0,656	0,756	0,843	0,936	1,030	1,124
14	0,328	0,437	0,546	0,656	0,769	0,874	0,984	1,092	1,201	1,312
15	0,351	0,468	0,585	0,702	0,822	0,936	1,053	1,170	1,287	1,404
16	0,374	0,498	0,624	0,748	0,872	0,996	1,122	1,248	1,373	1,496
18	0,421	0,562	0,702	0,842	0,982	1,124	1,263	1,404	1,545	1,684
20	0,468	0,624	0,780	0,936	1,092	1,248	1,404	1,560	1,716	1,872
22	0,514	0,687	0,858	1,028	1,202	1,374	1,542	1,716	1,888	2,056
24	0,562	0,749	0,936	1,124	1,312	1,498	1,686	1,872	2,060	2,248
25	0,585	0,780	0,975	1,170	1,366	1,560	1,755	1,950	2,145	2,340
26	0,608	0,812	1,014	1,216	1,420	1,624	1,824	2,028	2,230	2,432
28	0,655	0,874	1,092	1,310	1,530	1,748	1,965	2,184	2,402	2,620
30	0,701	0,937	1,170	1,402	1,640	1,874	2,103	2,340	2,574	2,804
35	0,819	1,092	1,365	1,638	1,911	2,184	2,457	2,730	3,003	3,276
40	0,936	1,248	1,560	1,872	2,104	2,496	2,808	3,120	3,432	3,746
45	1,053	1,405	1,755	2,106	2,466	2,810	3,159	3,510	3,861	4,212
50	1,170	1,561	1,950	2,340	2,830	3,122	3,510	3,900	4,290	4,680
55	1,287	1,717	2,145	2,574	3,003	3,434	3,861	4,290	4,719	5,148
60	1,404	1,873	2,340	2,808	3,276	3,746	4,212	4,680	5,148	5,616
65	1,521	2,029	2,535	3,042	3,549	4,058	4,563	5,070	5,577	6,084
70	1,638	2,185	2,730	3,276	3,822	4,370	4,914	5,460	6,006	6,552
75	1,755	2,341	2,925	3,510	4,098	4,682	5,265	5,850	6,435	7,010
80	1,872	2,497	3,120	3,744	4,368	4,994	5,616	6,240	6,864	7,488
85	1,989	2,653	3,315	3,978	4,641	5,306	5,967	6,630	7,293	7,956
90	2,106	2,809	3,510	4,212	4,914	5,618	6,318	7,020	7,622	8,424
95	2,223	2,965	3,705	4,446	5,187	5,930	6,669	7,410	8,051	8,892
100	2,340	3,121	3,900	4,680	5,460	6,242	7,020	7,800	8,580	9,360
110	2,574	3,433	4,290	5,148	6,006	6,866	7,722	8,580	9,438	10,29
120	2,808	3,745	4,680	5,616	6,552	7,490	8,424	9,360	10,30	10,43
130	3,042	4,057	5,070	6,084	7,098	8,114	9,126	10,14	11,15	12,17
140	3,276	4,369	5,460	6,552	7,660	8,738	9,828	10,92	12,01	13,10
150	3,510	4,651	5,850	7,020	8,220	9,362	10,53	11,70	12,87	14,04

per linear meter in kilogrammes.

Thickness in millimeters

Breadth in millim.	13	14	15	16	17	18	19	20	25	30
10	1,014	1,092	1,170	1,248	1,326	1,404	1,482	1,560	1,950	3,340
12	1,217	1,310	1,400	1,512	1,591	1,686	1,778	1,890	2,340	2,810
14	1,420	1,528	1,640	1,748	1,856	1,968	2,074	2,185	2,730	3,280
15	1,521	1,637	1,760	1,872	1,989	2,106	2,223	2,340	2,930	3,510
16	1,622	1,746	1,870	1,992	2,122	2,244	2,371	2,490	3,120	3,740
18	1,825	1,965	2,110	2,248	2,387	2,526	2,667	2,810	3,510	4,210
20	2,028	2,184	2,340	2,496	2,652	2,808	2,964	3,120	3,900	4,680
22	2,231	2,402	2,570	2,748	2,917	3,084	3,260	3,435	4,290	5,140
24	2,434	2,620	2,808	2,996	3,182	3,372	3,556	3,745	4,680	5,620
25	2,535	2,729	2,930	3,120	3,315	3,510	3,705	3,900	4,880	5,850
26	2,636	2,838	3,040	3,248	3,448	3,648	3,853	4,060	5,070	6,080
28	2,839	3,057	3.288	3,496	3,713	3,930	4,149	4,370	5,460	6,550
30	3,042	3,276	3,510	3,748	3,978	4,206	4,446	4,685	5,850	7,010
35	3,549	3,822	4,100	4,368	4,641	4,914	5,187	5,460	6,820	8,190
40	4,056	4,368	4.680	4,992	5,284	5,616	5,928	6,240	7,800	9,360
45	4,563	4,914	5,270	5,620	5,947	6,318	6,669	7,025	8,780	10,53
50	5,070	5,460	5,830	6,244	6,630	7,020	7,410	7,850	9,750	11,70
55	5,577	6,006	6,440	6,868	7,293	7,722	8,151	8,585	10,72	12,87
60	6,084	6,552	7,020	7,492	7,956	8,424	8,892	9,364	11,70	14,04
65	6,591	7,102	7,610	8,116	8,619	9,126	9,633	10,14	12,68	15,21
70	7,098	7,644	8,190	8,740	9,282	9,828	10,37	10,92	13,65	16,38
75	7,605	8,190	8,770	9,364	9,945	10,53	11,11	11,70	14,63	17,55
80	8,112	8,736	9,360	9,988	10,60	11,23	11,85	12,48	15,60	18,72
85	8,619	9,282	9,950	10,61	11,26	11,93	12,59	13,26	16,58	19,89
90	9,126	9,828	10,53	11,23	11,93	12,63	13,18	14,04	17,55	21,06
95	9,633	10,37	11,12	11,86	12,59	13,34	13,82	14,82	18,53	22,23
100	10,14	10,92	11,70	12,48	13,26	14,04	14,82	15,60	19,50	23,40
110	11,15	12,00	12,87	13,73	14,59	15,44	16,30	17,16	21,45	25,74
120	12,17	13,10	14,04	14,98	15,91	16,85	17,78	18,72	23,40	28,08
130	13,18	14,19	15,21	16,23	17,24	18,25	19,26	20,28	25,35	30,42
140	14,20	15,28	16,38	17,47	18,56	19,65	20,74	21,84	27,30	32,76
150	15,21	16,37	17,55	18,72	19,89	21,06	22,23	23,40	29,25	35,10

A LIST OF BOOKS ON
ORNAMENT, & DECORATION, ETC.

Published and Sold by
B. T. BATSFORD, 94 HIGH HOLBORN, LONDON W. C.

Mr. Lewis F. Day's Text Books of Ornamental Design.

Approved by the Science and Art Department.

Specially adapted for Use in Art Schools, and fully Illustrated.

SOME PRINCIPLES OF EVERY-DAY ART.

Introductory Chapters on the Arts not Fine. Forming a Prefatory Volume to the Series. Second Edition, revised, with 70 Illustrations. Crown 8vo, cloth Price 3/6.

"Authoritative as coming from a writer whose mastery of the subjects is not to be disputed, and who is generous in imparting the knowledge he acquired with difficulty. Mr. Day has taken much trouble with the new edition." — *The Architect.*

THE ANATOMY OF PATTERN.

Fourth Edition. Revised and Enlarged With 41 full-page Illustrations. Crown 8vo, cloth Price 3/6.

"A lucid analysis of repeated ornament. . . . . . A pre-eminently useful book." — *The Studio.*

THE PLANNING OF ORNAMENT.

Third Edition Further Revised. With 41 full-page Illustrations mostly re-drawn. Crown 8vo, cloth Price 3/6.

"Contains many apt and well-drawn illustrations, and is a highly comprehensive, compact, and intelligent treatise. It is a capital little book, from which no student can avoid gaining a good deal." — *The Athenæum.*

THE APPLICATION OF ORNAMENT.

Third Edition. With 48 full-page Illustrations. Crown 8vo, cloth Price 3/6.

"A most worthy supplement to Mr. Day's former works, and a distinct gain to the Art Student who has already applied his art knowledge in a practical manner, or who hopes ye to do so." — *Science and Art.*

ORNAMENTAL DESIGN.

Comprising the above Three Works, handsomely bound in one volume, cloth gilt, gilt top. Price 10/6.

NATURE IN ORNAMENT.

With 123 full-page Plates and 192 Illustrations in the Text. Third Edition (Third thousand) with a copious index. Thick Crown 8vo, cloth richly gilt, Price 12/6.

"A book more beautiful for its Illustrations or one more helpful to Students of Art can hardly be imagined." — *The Queen.*

"The Treatise should be in the hands of every Student of Ornamental Design. It is profusely and admirably illustrated, and well printed." — *The Magazine of Art.*

ALPHABETS.

For the use of Architects, Artists, Draughtsmen and Art Workers. Compiled by LEWIS F. DAY. With numerous illustrations of ancient and modern examples. *[In preparation.*

A HANDBOOK OF ORNAMENT.

With 300 Plates, containing about 3,000 Illustrations of the Elements and the application of Decoration to objects, by F. S. MEYER, Professor at the School of Applied Art, Karlsruhe. Second English Edition, revised by HUGH STANNUS F. R. I. B. A., Lecturer on Applied Art at the National Art Schools, South Kensington. Thick 8vo, cloth gilt, gilt top Price 12/6.

"A Library, a Museum, an Encyclopædia, and an Art School in one. To rival it as a book of reference, one must fill a bookcase. The quality of the drawings is unusually high, the choice of examples is singularly good. . . . . The text is well digested. . . . . The Work is practically an epitome of a hundred Works on Design." — *The Studio*.

"The author's acquaintance with ornament amazes, and his three thousand subjects are gleaned from the finest which the world affords. As a treasury of ornament drawn to scale in all styles, and derived from genuine concrete objects, we have nothing in England which will not appear as poverty-stricken as compared with Professor Meyer's Book." — *The Architect*.

"The book is a mine of wealth even to an ordinary reader, while to the Student of Art and Archæology it is simply indispensable as a reference book. We know of no one work of its kind that approaches it for comprehensiveness and historical accuracy." — *Science and Art*.

A HANDBOOK OF ART SMITHING.

For the use of Practical Smiths, Designers and others, and in Art and Technical Schools. By F. S. MEYER, Author of "A Handbook of Ornament". Translated from the Second German Edition. With an Introduction to the English Edition by J. STARKIE GARDNER. Containing 214 illustrations of Domestic and Architectural Wrought Iron work of the Medieval and Renaissance Periods. 8vo, cloth. Price 6/—.

ENGLISH MEDIÆVAL FOLIAGE AND COLOURED DECORATION.

By JAS. K. COLLING. A Series of Examples taken from Buildings of the Twelfth to the Fifteenth Century. *76 lithographic plates, and 79 woodcut illustrations*, with Text. Royal 4to, cloth, gilt top Price 18/—.

"The Author's book is one which it is a pleasure to recommend. He has done his work as well as it could be done, and we trust he may be encouraged by the success of this book to take up some other details of Mediæval work." — *The Architect*.

DETAILS OF GOTHIC WOOD CARVING.

Being a series of drawings from original work of the XIVth, and XVth centuries. By Franklyn A. CRALLAN. Late Instructor in Wood Carving, Municipal Technical College, Derby

This work will consist of 34 plates, of which two will be double, reproduced by Photo-Tint process from the original full-sized drawings made by the author, chiefly from Churches in Derbyshire, Norfolk, and Surrey, also from the Cathedral Churches of Ely, Peterborough, and Lincoln, and the Abbey Church at Westminster.

The size of the plates will average 12 by $8^1/_2$ in., printed on paper 15 by 11 in., and they will be accompanied by short descriptive text.

THE PRICE OF THE WORK, IN SUITABLE PORTFOLIO, WILL BE 21 S. TO SUBSCRIBERS ONLY. UPON PUBLICATION THE PRICE WILL BE RAISED TO 28 S. [IN PREPARATION.

SPECIMENS OF ANTIQUE CARVED FURNITURE AND WOODWORK.

Measured and drawn by ARTHUR MARSHALL, A. R. I. B. A., *50 Photo-lithographed plates* of English Examples, with Sections, Mouldings, and Details, and Descriptive Text containing numerous other details. Folio, half-bound in calf Price 2/2/— net.

This Work is the most complete and valuable that has been issued illustrating old English Carved Woodwork. It is filled with useful Details drawn to a large scale, so as to make it invaluable to Woodcarvers.

"To all who are interested in the beautiful and useful art of wood-carving, the book under our notice will be very welcome. It is a folio with clear illustrations, fit to use as working copies, of many authentic specimens of antique carved furniture and woodwork." *Saturday Review*.

HEPPELWHITE'S CABINETMAKER AND UPHOLSTERER'S GUIDE.

A complete facsimile reproduction of this rare work containing over 300 charming designs on 127 plates, of every article of Household Furniture. By A. HEPPELWHITE & Co. Small folio, halfbound (1789). Price 2/10/— *[In preparation.*

"Their tea-caddies, tea-trays, tops of card-tables and dressing-tables are most charming examples of beautiful design and arrangement. . . . The book taken as a whole is useful and modest, and nearly always quite practicable, so that among their 300 designs there are scarcely twenty which might not, with advantage, be reproduced." — *J. Aldam Heaton in "Furniture and Decoration of the 18th Century."*

SHERATON'S CABINET-MAKER AND UPHOLSTERERS' DRAWING BOOK.

With the rare Appendix and Accompaniment. A complete Reproduction of this scarce and costly Volume, with *122 plates* of Examples of all kinds of Furniture and Details designed by THOMAS SHERATON, thick 4to, cloth, gilt top (1802) Price 2/10/— net. *[Just published.*

CHIPPENDALE'S "THE GENTLEMEN AND CABINET-MAKER'S DIRECTOR."

A complete facsimile of the 3rd and rarest Edition, of this celebrated work containing *200 plates* of Designs of Chairs, Sofas, Beds and Couches, Tables, Library Book Cases, Clock Cases, Stove Grates, &c. &c. Folio, half-calf (1762) Price 4/4/— net.

"This should be consulted by every Student of Furniture. It is Chippendale's Work reprinted, contains his own Designs, and shows the manner in which the 18th Century Cabinet-maker went about his work." — C. R. ASHBEE, *Hon. Director of the Guild and School of Handicraft.*

FURNITURE AND DECORATION IN ENGLAND DURING THE XVIIITH CENTURY.

Facsimile Reproductions of the Choicest Examples from the Works of CHIPPENDALE, ADAM, RICHARDSON, HEPPELWHITE, SHERATON, PERGOLESI, and others. Selected and Described by JOHN ALDAM HEATON. Containing 200 folio plates, with descriptions. Price in Four Cloth Portfolios 6/—/— net., or Bound in Four Volumes in cloth 7/—/— net.

"The Furniture of the latter half of the last century has, of late, so commonly come to be regarded as the best the world has yet produced, and the cost of obtaining a complete set of the illustrations is so great, that it has seemed advisable to now publish a set of reproductions of the most admired and the most useful.

"Mr. Aldam Heaton's name is that of a real authority on the subject, and, as the result proves, no better judge could have been chosen to direct such a publication. His method has been to go through the series of some twenty publications, published between 1740 and 1800, and to select whatever designs appeared to him the most valuable. The rarity of the original volumes, and the very high prices which they realise at sales, put them practically out of the reach of the architect and cabinet-maker of to-day, so that, if any use is to be made of them it must be through some of the reproductions which modern process printing makes possible." — *The Times.*

DECORATIVE DESIGNS.

By ROBERT ADAM. A Series of Six fine large plates of Ceilings, Scroll Ornament, &c., being transfers taken from the original drawings in the Soane Museum by R. CHARLES. Oblong 4to, folded in wrapper. Price 2/6 net.

DECORATIVE WROUGHT IRON WORK OF THE 17TH AND 18TH CENTURIES.

16 large photo-lithographic plates, containing 70 English examples of Measured Drawings, by D. J. EBBETTS, of large and small Gates, Screens, Grilles, Panels, Balustrading, &c. Folio, boards, cloth back. Price 12/6.

A HISTORY OF DESIGN IN PAINTED GLASS.

By N. H. J. WESTLAKE, F. S. A. Containing 467 beautiful illustrations. 4 vols, small folio, cloth. Price 5/10/—

The arrangement of this valuable work, the most complete yet attempted on its subject, is as follows, and the volumes with the exception of volume 1 may be had separately at the prices affixed: —

VOLUME I. Part I. From the Earliest Examples until the End of the Twelfth Century. Part II. Single Figures and Simple Compositions of the Thirteenth Century. Part III. Medallion and Grisaille Windows of the Thirteenth Century.

VOLUME II Part IV. Introduction, and English Fourteenth Century Work. Part V. French, German, and Italian Fourteenth Century Work, Grisaille, and Quarries of the Fourteenth Century. Price 1/1/—

VOLUME III. Part VI. Introduction, and Preface to the Fifteenth Century; English Figure Windows of the Fifteenth Century. Part VII. The Jesse Tree, English Subject and Windows of the Fifteenth Century. Part VIII. French Subject and Figure Windows of the Fifthcenth Century, German and Italian. Part IX. Heraldry, Roundlets, Monograms, Quarries, &c. Price 1/10/—.

VOLUME IV. Preface and Introduction to the Sixteenth Century. English and Foreign Painted Glass in England of the Sixteenth and Seventeenth Centuries. French, Flemish, Dutch, German, Italian and Spanish Painted Glass of the Sixteenth and Seventeenth Centuries. The Jesse Tree, Ornamental Windows, &c. Index. Price 1/14/—.

ANCIENT WOOD & IRONWORK IN CAMBRIDGE.

By W. B. REDFARN, the Letterpress by JOHN WILLIS CLARK, *29 folio lithographed plates* drawn to a good scale. Cloth, gilt, a handsome volume. Price 10/6.

This Work, giving an interesting and useful series of Examples, is but little known. The few copies existing are offered for the present at the reduced price as above.

REMAINS OF ECCLESIASTICAL WOOD-WORK.

A Series of Examples of Stalls, Screens, Book-Boards, Roofs, Pulpits, &c., containing 21 plates beautifully engraved on Copper, from drawings by T. TALBOT BURY, Archt. 4to halfbound Price 10/6. First published at 25 s.

PROGRESSIVE STUDIES & OTHER DESIGNS FOR WOOD-CARVERS

in various Styles, Sheet No. I., consisting of Progressive Studies, Elementary and Advanced. for Class Teaching. Sheets Nos. II., III., and IV., including a variety of objects suitable for Wood-Carving; Table, Spinning Chair, Koran Stand, Frames, Bellows, &c, by Miss E. R. PLOWDEN, with a Preface by Miss ROWE. The Sheets folded in royal 8vo, portfolio Price 5/— net.

HINTS ON WOOD CARVING.

For Beginners, by ELEANOR ROWE, with a Preface by J. H. POLLEN. 4th Edition, revised and enlarged, *illustrated*. 8vo, sewed Price 1/—, or cloth Price 1/6.

HINTS ON CHIP CARVING.

(Class Teaching and other Northern Styles), by ELEANOR ROWE. *40 illustrations.* 8vo, sewed Price 1/—, or bound in cloth Price 1/6.

Miss ROWE is at the School of Art Woodcarving, and her books are published under the sanction of the Science and Art Department.

PLASTERING, PLAIN AND DECORATIVE.

A Practical Treatise on the Art and Craft of Plastering. Including full descriptions of the various Tools, Materials Processes and Appliances employed. With over 50 full-page plates, and about 300 smaller illustrations in the text. By WILLIAM MILLAR. With an Introduction, treating of the History of the Art, with numerous examples. 4to, cloth Price 18/— net. *[In preparation.*

FLAT ORNAMENT.

A Pattern Book for Designers of Textiles, Embroideries, Wall Papers, Inlays &c &c. *150 plates,* some printed in Colours. exhibiting upwards of 500 Examples of Textiles, Embroideries, Paper Hangings, Tile Pavements, Intarsia Work, Tapestries, Bookbindings, Surface Ornaments from Buildings, &c. &c., collected from various Museums, Churches, Mosques, &c. &c., with some Original Designs for Textile and other Ornament, by Drs. FISCHBACH, GIRAUD, and others. Imperial 4to boards, cloth back Price 25/—.

JAPANESE ORNAMENT AND DESIGN.

A Grammar of, *illustrated by 65 plates,* many in Colours and Gold, representing all Classes of Natural and Conventional Forms, drawn from the Originals; with Introductory, Descriptive, and Analytical Text. By T. W. CUTLER, F. R. I. B. A. Imperial 4to. in elegant cloth binding. Price 2/10/—

"The work is one which is almost indispensable to every decorative designer, including as it does specimens of the very best Japanese Art applied to the ornamentation of various materials." — *The Queen.*

"The beauty of Mr. Cutler's folio pages of illustrations, the systematic order of their arrangement, and the production of a certain proportion of the plates in the rich gold and colours of the original Japanese embroidery give instructive value as well as pictorial charm to this *beautiful work.*" — *Edinburgh Review.*

"Mr. Cutler's 'Grammar of Japanese Ornament and Design' is one of the most elaborate and comprehensive works on the subject that has yet been published. It presents the art of the far East in all its beauty and its bewildering strangeness." — *The Times.*

JAPANESE ART BOOKS (NATIVE PRINTED).

A Charming Series of Studies of Birds in most characteristic and life-like attitudes, with appropriate foliage and flowers, by the celebrated Japanese artist, BAIREI KONO. Three books, 8vo, containing 108 pages *of highly artistic and decorative illustrations printed in tints.* These books are of the greatest value to Artists, Screen and China Painters, Decorators, and Designers in all branches of Art Manufacture, and of much interest to the admirers of Japanese Art. Price 3/6 each book, or the set of three 10/— net.

"The books consist of characteristic studies of Birds in every variety of attitude, skimming through the air, proudly, defiantly, or saucily perched upon light graceful boughs, swimming in search of food, or devouring the frogs and fish which have just been caught. For variety both of bird and tree form the designs are inimitable. . . . . . The pride of the peacock, as expressed in the haughty turn of the head, and disdainful eye, the wisdom of the owl, the gaiety of the small birds on the wing, and the eager and expectant look of those about to pounce on tender morsels, all speak for themselves, and speak highly for the subtle manipulative skill of the artist." — *The Graphic.*

"In attitude and gesture and expression, these Birds, whether perching or soaring, swooping or brooding, are admirable." — *Magazine of Art.*

STUDIES OF JAPANESE BIRDS AND FLOWERS.

By WATANABE SEITEI, the acknowledged leading living artist in Japan. Three books, 8vo, containing numerous artistic sketches in various tints. Price 10s net.

"Contain a wealth of exquisite xylographic impressions, which cannot be beaten by any European attempts." — *The Studio.*

Two Exhibitions of the Artist's Drawings have been held at the Japanese Gallery, and the Drawings have fetched high prices.

JAPANESE ENCYCLOPÆDIA OF DESIGN (NATIVE PRINTED).

BOOK I. — Containing over 1,500 engraved, curious, and most ingenious *Geometric Patterns* of Circles, Medallions, &c., comprising Conventional Details of Plants, Flowers, Leaves, Petals, also Birds, Fans, Animals, Key Patterns, &c., &c. Oblong 12mo. Price 2/— net.

BOOK II. — Containing over 600 most original and effective Designs for Diaper Ornament, giving the base lines to the design, also artistic Miniature Picturesque Sketches. Oblong 12mo. Price 2/— net.

These books exhibit the varied charm and originality of conception of Japanese Ornament, and form an inexhaustible field of design.

A HISTORY OF ARCHITECTURE.

For the Student, Craftsman, and Amateur. Being a Comparative View of the Historical Styles from the Earliest period. By PROFESSOR BANISTER FLETCHER, F.R.I.B.A., and BANISTER F. FLETCHER, A.R.I.B.A. With 115 plates printed in collotype from large photographs, and other illustrations in the text. Thick crown 8vo, cloth gilt. Price 12/6. [*Just published.*

"We shall be amazed if it is not immediately recognized and adopted as *par excellence* the student's manual of the history of architecture."—*Architect.*

"Concisely written and profusely illustrated by plates of all the typical buildings of each country and period. The authors have written a book which appears to fulfil the necessary qualifications of being of moderate compass, and giving in a complete and classified form the results of the latest researches of architectural historians and archæologists, and at the same time furnishing the student with a comparative and analytical view of the subject. . . . It will fill a void in our literature."—*Building News.*

"The architectural student in search of any particular fact will readily find it in this most methodical work . . . . as complete as it well can be."—*Times.*

"A handy and compact volume . . . admirable alike in plan and execution. It is scarcely possible, within the compass of 300 small pages to carry completeness further."—*National Observer.*

RENAISSANCE ARCHITECTURE AND ORNAMENT IN SPAIN.

A Series of Examples selected from the purest executed between the years 1500—1560, by ANDREW N. PRENTICE, Architect. 60 *plates*, reproduced by Photo-lithography and Photo Process, from the drawings made by the author during recent visits to Spain. With the exception of five or six perspective views, the plates consist of geometrical drawings carefully prepared on the spot from measurements, and include details of Façades, Patios, Staircases, Doors, Windows, Ceilings, Brackets, Capitals, and other details in Stone and Wood, together with examples of Iron Screens, Balconies and other specimens of Metal Work &c. &c. A Short Descriptive Text is added. Folio, handsomely bound in cloth gilt Price 2/10/—.

"There is no doubt that of all the books which have Spanish buildings for subject, Mr. Prentice's will be found the most fascinating by an English Architect."—*Architect.*

"For the drawing and production of this book one can have no words but praise. . . . It is a pleasure to have so good a record of such admirable Architectural Drawing, free, firm, and delicate."—*British Architect.*

ITALIAN RENAISSANCE DETAIL AND ORNAMENT.

Drawn by G. J. OAKESHOTT, Architect, A.R.I.B.A., "Building News" and "Aldwinckle" Travelling Student. 40 *lithographic plates*, royal folio, strongly bound in cloth. Price 1/12/—

"One of the best produced and most useful of professional illustrated works we have seen for some time is the 'Detail and Ornament of the Italian Renaissance', by Mr. George J. Oakeshott. The series of forty plates includes some exquisite examples of Renaissance art illustrated almost entirely in the form of geometrical drawings. We have nothing but praise for this work, which is throughout carefully and conscientiously carried out. The author deserves commendation for his choice of subjects, no less than for his illustration of them."—*British Architect.*

THE ORDERS OF ARCHITECTURE—GREEK, ROMAN, & ITALIAN.

Selected from Normand's Parallels and other Authorities. Edited with Notes by R. PHENÉ SPIERS. Second Edition, with four new plates, including one of Greek Mouldings, by R. W. SCHULTZ. 24 *plates*, imperial 4to, cloth Price 10/6.

"A most useful work for architectural students . . . . Mr. Spiers has done excellent service in editing this work, and his notes on the plates are very appropriate and useful."—*British Architect.*

"The book is certainly a thoroughly practical work, bringing careful representations of the Classical Orders readily and clearly within the modest reach of most students, and cannot fail to be useful as a book of reference."—*Building News.*

"Should be considered as an indispensable possession by all students of architecture."—*Architect.*

B. T. BATSFORD,
ARCHITECTURAL & ART BOOKSELLER & PUBLISHER.
94, HIGH HOLBORN, **LONDON.**

www.ingramcontent.com/pod-product-compliance
Lightning Source LLC
Chambersburg PA
CBHW021846230426
43669CB00008B/1095